CROP PRODUCTION EQUIPMENT

CROP
PRODUCTION
EQUIPMENT

H. T. LOVEGROVE, N.D.Agr.E., A.M.I.Agr.E.

HUTCHINSON EDUCATIONAL

HUTCHINSON EDUCATIONAL LTD
3 Fitzroy Square, London W1

London Melbourne Sydney Auckland
Wellington Johannesburg Cape Town
and agencies throughout the world

First published 1968
Second impression 1971
Third impression 1973

Printed in Great Britain by litho on smooth wove paper
by Anchor Press, and bound by Wm. Brendon,
both of Tiptree, Essex

ISBN 0 09 085390 3

To my wife Sylvia, and daughters Ruth and Kathryn, whose tolerance and encouragement during my many hours of preoccupation enabled me to complete the writing and illustrating of the text.

CONTENTS

CONTENTS

PLATES

PREFACE

The cost and complexity of farm machinery increases year by year, and in order to justify the capital investment involved, it is essential that operators should be able to exploit to the full each machine's capacity.

Efficient operation calls for a sound knowledge of the working principles of machines and an understanding of the basic husbandry objectives of each operation.

This book sets out to unite these two factors by explaining the relationship between machines, soils and crops.

In view of the size of the subject, I thought it preferable to deal with one section of farm mechanisation in depth, rather than to produce a cursory treatment of the whole subject. It was my opinion that crop production equipment should take priority. Although this excludes tractors as a subject in its own right, they are referred to wherever necessary in relation to field machinery.

Each machine is dealt with under the following headings: Aims and objects of the process; Construction and working principles; Alternative types of machine (for any one operation or set of operations); Operational adjustments; Field procedure; Maintenance; Safety precautions.

The essentially practical approach should be of particular appeal to farm apprentices, trainees and students of all levels. I hope experienced tractor drivers and farmers themselves will find the more advanced material a valuable source of help and guidance in fault analysis and choice of machines.

The scope of the book embraces the crop production equipment content of the following schemes and syllabuses: City and Guilds Farm Machinery Stages I and II (267 and 270), Crop Husbandry Stage II (268), Animal Husbandry Stage II (269), Union of Educational Institutes (7) NCA, National Proficiency Tests (8), NDA.

I have intentionally over-simplified some of the line drawings to lay stress on key points. Film strips are available.

Horsham, 1968   H. T. LOVEGROVE

ACKNOWLEDGEMENTS

My grateful thanks are due to Charles Boyce and Ben Walters, on whose farms I acquired much valuable experience in crop husbandry techniques, and was awakened to the enormous scope in agriculture for my earlier engineering training; Colin Foulkes, who took the risk of engaging me on my first teaching appointment which gave me seven happy years in Norfolk; Don Park, Principal of the Shropshire Farm Institute, who first prompted me to write this book, and on whose staff I spent four most stimulating years; my friends and colleagues on the staff of the Wolverhampton Technical Teachers College who goaded me into finishing the work; Pat Wootton and Jennifer Bushell who, with my wife, typed the manuscript; Arthur Lock and Ivor Watkins, who assure me of the sale of at least two copies of this book!

I would also like to thank the following firms for permission to reproduce illustrations: E. Allman Ltd., Blanch-Lely Ltd., Fisons Fertilisers Ltd., Massey-Ferguson Ltd., Ransomes, Sims & Jefferies Ltd., John Salmon Ltd., Steelfab Ltd., Ernest A. Webb Ltd., Wellington Journal and Shrewsbury News, John Wilder Ltd.

1 PLOUGHS

Ever since man started to grow his own food the plough has been his main tillage implement. The first crude wooden ploughs were drawn through the soil to loosen it a little before the seeds were sown. The modern plough has developed by an extremely slow process, helped in recent years by the application of science to farm machine design.

Now that farmers are better informed in matters of crop husbandry, ploughing is known to be a much more important and involved procedure than the mere breaking down of the soil. The production of the ideal soil environment for the germination of seeds, and for the development of crops, entails a number of processes, as follows:

PRODUCTION OF TILTH

By exposing the soil in preparation for the pulverising effects of implements the plough plays a vital part in the production of *tilth*, the fine soil texture necessary to ensure close contact between the soil and the seeds and subsequent plant roots.

CONTROL OF WEEDS

By the inversion of top soil, weeds are severed and their surface foliage is buried.

AERATION

The breaking down of the soil facilitates the entry and free circulation of air. This, in turn, promotes the activity of micro-organisms in the soil, which make food available for the plants.

EXPOSURE TO WEATHER

By the exposure of fresh soil to the atmosphere, and to the changes of temperature and humidity brought about by the wind, sun, rain and frost, further aeration and tilth production is achieved.

BURIAL OF RESIDUES

When crop residues and farmyard manure are ploughed into the soil they decompose and provide humus which improves the crumb structure of the soil, i.e. the crumb-like grouping of fine soil particles, necessary for good circulation of air and moisture. Also, the material ploughed in supplements the plant foods.

DRAINAGE

By the opening and loosening of the top soil, excess surface water is able to filter through to the subsoil and so, eventually, into the field-drainage system.

The relative importance of the various processes mentioned may vary between one field and another, according to the type and condition of the soil to be ploughed, the time of year, and the requirements of the intended crop.

Construction and working principles of mouldboard ploughs

In order to appreciate how all these processes can be achieved in

2

the single operation of ploughing, the action of each of the soil-working parts of a typical mouldboard plough must be understood. The large majority of ploughs in Britain are mouldboard ploughs, and this chapter is concerned mainly with this type. A short description of disc, chisel, and mole ploughs appears at the end of the chapter. Fig 1 indicates the parts of a mouldboard plough and Plate 1 shows the plough at work.

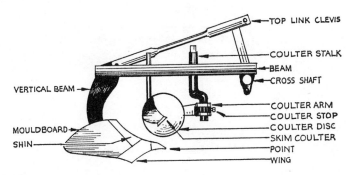

1 Single furrow plough

DISC COULTER

The disc coulter is a hard steel disc, usually about 16 in in diameter, which is free to rotate on a spindle. The spindle bearing is located in a single or forked coulter arm which is free to swivel within limits about its supporting vertical stalk. The function of the coulter is to make the vertical cut which severs the furrow slice from the unploughed land. The various adjustments which can be made to the coulter to suit working conditions will be discussed under *Operational adjustments*.

KNIFE COULTER

The knife coulter is simply an alternative to the disc type just described (Fig 2). It is supplied as standard equipment on many reversible ploughs, and its advantage over the disc type is that there is less likelihood of the plough *choking*, especially in land

3

with troublesome roots, stones or trash; it is also advantageous in hard, dry soil conditions where disc coulters tend to ride on the surface and so reduce share penetration.

Further virtues of the knife coulter are its simplicity, in that it has no moving parts, and its lightness by comparison with the

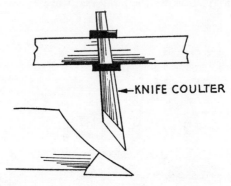

←—KNIFE COULTER

2 Knife coulter

disc-coulter assembly. Weight is an important consideration where reversible ploughs are concerned, because of the problem of weight overhang at the rear of the tractor. It must be said, however, that the knife coulter imposes a higher draught on the tractor than does the disc type.

SKIM COULTER

Sometimes referred to as the *skimmer* or *jointer*, a skim coulter is positioned close to the side of the line taken by the disc or knife coulter. Its object is to chamfer off a narrow strip of surface vegetation from the left-hand corner of the furrow slice before it is inverted to discourage the growth of weeds between the furrows and to seal the furrow joints—hence the name *jointer*. The skimming of trash is deflected from the surface of the furrow slice into the bottom of the adjacent furrow trench (Figs 29 and 31).

4

SHARE

The share makes the horizontal cut which severs the slice of top soil from the subsoil. It is made either of chilled cast iron or of steel, and in one of a variety of patterns. Steel is less brittle than chilled iron and, therefore, permits a thinner section to be used, giving cleaner cutting and reduced cutting resistance. The more traditional socket-fitting share (Fig 3) is designed with a slight

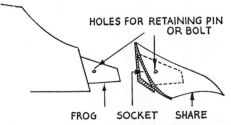

HOLES FOR RETAINING PIN OR BOLT

FROG   SOCKET   SHARE

3 One-piece socket fitting share (chilled cast iron or steel)

downward and lateral inclination of the point to increase penetration both downwards and sideways towards the unploughed land. These features, known as the *suction* and the *lead to land* of the share, respectively, become eliminated as the share wears (Fig 4). On some ploughs the share is replaced by a two-piece, or

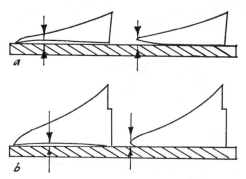

a

b

4 New and worn shares against straight-edge: *a* Elevation, *b* Plan view

5

even three-piece, assembly (Fig 5), the parts being known as the *point*, the *wing*, and the *shin* (which is sometimes known as the *cutter* or *comb*).

Some shares are produced with a harder underside for self-sharpening effect.

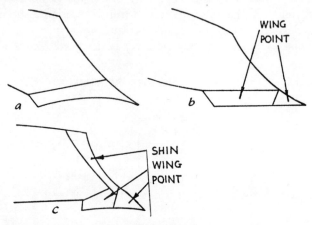

5 Types of share. *a* Single-piece steel plate, *b* Two-piece (socket fitting point), *c* Three-piece assembly

MOULDBOARD

Commonly constructed as a steel laminate with a layer of hard-wearing steel on its friction surface, the mouldboard turns the furrow slice through approximately 130° to lay it against the preceding slice. The design of the mouldboard is important for its *scouring* qualities, i.e. its freedom from soil adhesion to the friction surface, and for the type of work that it produces; the implications of the latter are discussed under *Types of plough*.

LANDSIDE PLATE

The slight angular arrangement of the mouldboards in relation to their direction of travel causes a side-thrust on the plough. A landside plate is fitted to each plough body on the opposite side to the mouldboard (Fig 6) to take the brunt of this pressure against

the furrow wall. On a multi-furrow plough the front and any intermediate landside plates are shorter than the rear one to give a
clearance between the bodies. On some ploughs an obliquely
mounted rear wheel is fitted to counter this side-thrust.

Some landside plates are designed to carry part of the plough's
rear-end weight, and these are usually reinforced on their underside with a steel or iron pad known as a *slade*.

FROG

The frog (or frame) is a robust steel or malleable iron casting to
which the share, mouldboard and landside plate are attached. In
some cases it is a positive fixture, in others a limited amount of

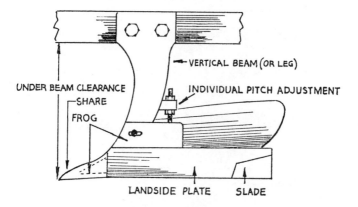

6 Landside of plough body

adjustment can be made in its relation to its supporting beam or
leg (Fig 6). This is usually referred to as *individual pitch adjustment*.

BODY

The term *body* (or *base*) applies to the complete assembly of soil-
working parts required for each furrow, comprising the frog,
share, mouldboard and landside plate. It is the design of this
assembly which is of the greatest significance when selecting a

7

plough, since it determines the type of work the plough will produce. The easy removal of a complete body to reduce draught in difficult conditions is also a valuable feature provided on some ploughs, and the ability to adjust the lateral spacing of the bodies is a further asset.

BEAMS

The beams form the basic structure of the plough. Sometimes they are curved to carry the plough bodies (Fig 7), in other cases they are horizontal and carry vertical legs to support the bodies (Fig 1).

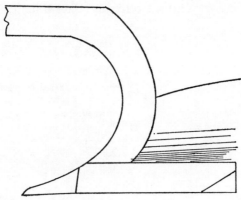

7 Curved plough beam

CROSS-SHAFT

The cross-shaft is a steel shaft positioned at the extreme front of the plough and arranged at right angles to the beams (Fig 1). It may be either above or below the beams, and on some ploughs either position is possible in order to modify the vertical line of draught to suit specific conditions. Each end of the cross-shaft is cranked and machined to provide attachment points for the tractor lower links. The cranks are diametrically opposed so that, as the shaft is rotated by the control lever provided, the plough may be angled to take either a wider or a narrower front furrow (Fig 8). The aim of this adjustment is shown in Fig 9.

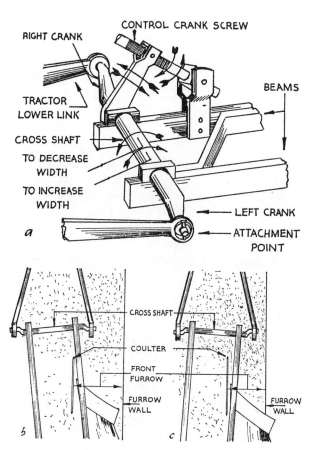

8 Front furrow width. *a* Adjustment—rotation of cross shaft
alters ploughing width of front furrow, *b* Maximum width,
c Minimum width. *Note* in *b* and *c* position of cross-shaft cranks

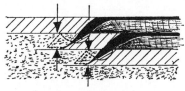

9 Front furrow width should be equal to those that follow

MECHANICAL PROTECTIVE DEVICES

Some ploughs have protective devices in the form of shear bolts or spring-release devices on the bodies and/or coulters, to relieve the impact of any obstruction in the soil.

Types of mouldboard plough

Mouldboard ploughs may be classified in a number of ways, e.g. type of body, mode of attachment to tractor, whether righthand or reversible action, or by the number of bodies it carries.

TYPE OF BODY

The type of body determines a plough's suitability for particular soil and working conditions. At the beginning of this chapter the various effects of plough action have been outlined. In order to achieve these in the widely varying working conditions that may be encountered, alternative pattern plough bodies are made. Until recently, these could be classified as *general-purpose*, *semi-digger*, or *digger* types, but developments in plough-body design have made this classification less distinctive. In order to increase versatility, the characteristics of many modern ploughs overlap the classification boundaries of earlier types, resulting in what might be called multi-purpose ploughs. The versatility of plough bodies is influenced by the following features:

Length and shape of mouldboard

Fig 10a shows a long, slow-turning mouldboard, and the well-defined unbroken furrows such a body would produce in clean-cutting cohesive soil. This is the result that should be aimed at when ploughing grassland and clover ley, where good packing of the furrow slices is important. It is also suitable for some winter ploughing, since the better the furrows are *set up*, the greater is the surface area of soil exposed to the weather.

Fig 10b shows a shorter mouldboard and the partially broken

work which results from its more rapid turning action on the furrow slice. This kind of work is generally suitable for ploughing when the soil is to be worked down immediately for drilling or planting. Some farmers, however, favour this type of body for the bulk of their work, since it leaves a more *open* finish.

Fig 10c shows the shattering effect a concave mouldboard has on the furrow slice caused by its abrupt forward-throwing action.

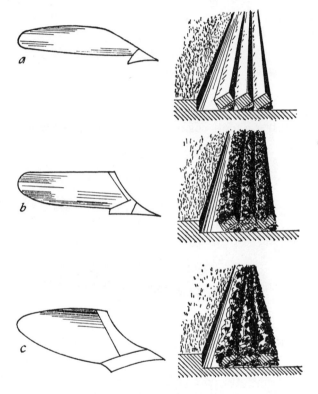

10 Types of body and examples of their work
a Relatively long, slow turning mouldboard. *Right* work produced in a cohesive soil—well defined, unbroken furrows. b Shorter, deeper mouldboard. *Right* work is more shattered. c Very short, concave mouldboard. *Right* its severe shearing action results in very broken furrows

This type of plough is widely used for deep ploughing and for the cross-ploughing of land previously ploughed at a shallower depth.

Ground speed of the plough can influence considerably the type of work produced by a particular pattern of mouldboard, and in order to increase working speeds without throwing and shattering the furrow slice unduly, specially designed mouldboards are being marketed.

To achieve scouring on sticky soils, good mouldboards are designed in such a way as to avoid areas of low soil pressure on the friction surface.

Bar-point bodies

Fig 11 illustrates a bar-point body, which is particularly advantageous in arduous conditions. A rigid or spring-loaded steel bar with a bevelled leading edge takes the brunt of the wear borne by

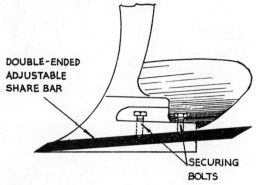

DOUBLE-ENDED
ADJUSTABLE
SHARE BAR

SECURING
BOLTS

11 Bar-point body

the share point of a normal body. If the body should strike an obstruction in hard, stony or rooty land, the likelihood of serious damage is small—especially on the spring-loaded types.

Under-beam clearance

Under-beam clearance is an important factor in a plough's

working-depth range. Fig 6 illustrates the dimension to which the term refers.

POINT-TO-POINT CLEARANCE

As illustrated in Fig 12, point-to-point clearance refers to the longitudinal distance between the bodies. Good point-to-point clearance is particularly valuable when ploughing in wet and trashy conditions, and for cross-ploughing where blockages are apt to occur. Unfortunately, increased point-to-point clearance means increased plough length, and this in turn means increased weight

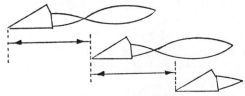

12 Point-to-point body clearance measurement

overhang at the rear of the tractor. As a result, this feature is less easily obtained on mounted ploughs than it was on the now obsolete trailed plough. In some cases—usually with reversible ploughs—the number of bodies a tractor can cope with is limited by this weight over-hang factor rather than by the pulling power of the tractor.

To reduce this imbalance, some makers recommend the fitting of ballast weights to the front wheels of the tractor, particularly on hilly land. Point-to-point clearances vary from 24 in to 40 in on different ploughs, although the majority fall in the range of 24 in to 36 in.

MODE OF ATTACHMENT TO TRACTOR

Trailed type

As its name implies, the trailed type of plough is simply carried on two or three wheels and is connected to the tractor by a draw pin. The arrangement of the drawbar assembly is critical if side draught and its attendant poor work and excessive strain are to be avoided. However, since relatively few trailed ploughs are now in use, explanation of drawbar arrangements is hardly justified.

Mounted type

The introduction of hydraulics on tractors as a means of auxiliary power has revolutionised the attachment and operation of ploughs as well as that of many other implements and machines. Three-point linkage attachment has brought such advantages as initial reduction in cost of the plough, easier transportation, simpler and more accurate control, better manœuvrability, reduced headland compaction, and improved traction by transferring the plough's weight to the tractor. By means of the three-point linkage attachment common to most makes of machine, the tractor and plough become virtually one unit.

Some ploughs are fitted with a land wheel. Others are totally suspended and dependent upon the tractor's hydraulic system (Fig 13). This results in the weight of the plough, and the soil it carries when in work, being transferred on to the rear wheels of

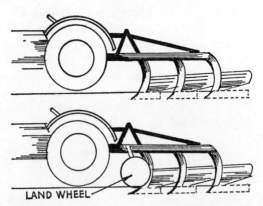

LAND WHEEL

13 Mounted ploughs: *Top* totally suspended, *Bottom* land wheel type

the tractor, so improving their adhesion (Fig 14). Furthermore, the *suck* of the plough—i.e. the penetration of the shares into the ground due to their angle of inclination—increases the downward pressure on the tractor wheels.

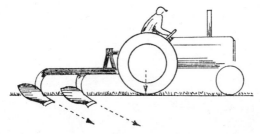

14 Tractor mounted wheel adhesion is increased by weight and downward *suck* of plough

Semi-mounted type

In the case of the semi-mounted type, the front of the plough is supported on the tractor drawbar or linkage, and the rear by castor wheels (Fig 15). This arrangement is sometimes used for

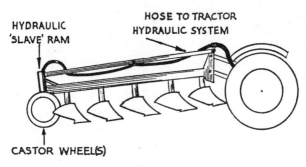

15 Semi-mounted plough with hydraulic tail lift equipment

the attachment of long or heavy ploughs which would otherwise impose an excessive load on the tractor's hydraulic system and upset the tractor's stability by excessive rear overhang.

RIGHT-HAND AND REVERSIBLE TYPES

For many years, most tractor ploughs have been constructed with right-hand bodies designed to throw the soil to the right (viewed from the rear of the plough). This arrangement, however, is rapidly giving place to the reversible plough which has both left-hand and right-hand bodies (Fig 16). By using these alternately, the outfit is

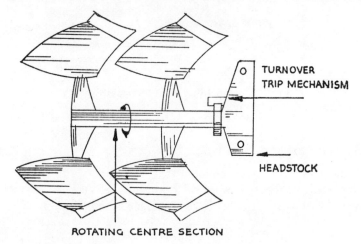

TURNOVER
TRIP MECHANISM

HEADSTOCK

ROTATING CENTRE SECTION

16 Reversible plough

able to progress across the field bout by bout laying all the furrows in one direction, hence its alternative, albeit paradoxical, name, *one-way* plough (Fig 17). The idea of one-way ploughing is by no means new—a number of reversible ploughs were manufactured for horse draught, for steam-ploughing tackle and for early tractors. The advantages over the fixed body, right-hand type are these: elimination of the time-consuming procedure of marking out fields and forming opening ridges and finishing furrows (see *Field procedure*), elimination of the idle running necessary when ploughing in lands, and production of a more level surface to the field because of the absence of openings and finishes. The latter is a most important factor where mechanical inter-row cultivations are intended. One-way ploughing also shows a distinct saving of both time and fuel.

In spite of these advantages, large-scale production of reversible ploughs was slow to develop. The fact that they are now produced in large numbers is mainly due to the simplification of plough construction resulting from the three-point linkage system of attachment and hydraulic control, and the availability of lighter materials of adequate strength and acceptable cost.

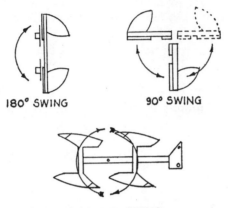

180° SWING   90° SWING

180° LONGITUDINAL SWING

17 Body changeover arrangements

Variations of detail exist, particularly in the design of the main supporting beams and the system employed to alternate the bodies. Tubular and skeleton frames have been introduced to overcome the weight-factor problem. In some ploughs, the section carrying the bodies rotates through 180° about the longitudinal axis, while on others a 90° swing is employed (Fig. 17). Other variations are longitudinal rotation of the bodies on a lateral pivot, and mounting of the bodies on separate beams designed to rise and fall independently. Plate 2 shows a reversible plough at work.

NUMBER OF BODIES

The prudent purchaser of a plough considers the capacity of his

17

tractor and the nature of the soil on his farm, since the extremities of soil types can effect a tractor's ploughing capacity by more than one furrow.

Use of an excessively large plough should be avoided for mechanical and economic reasons. Most ploughs in general use today are of two- or three-furrow capacity, although on some of the lighter soils four and even five furrows are handled quite ably by medium-sized tractors. In work requiring a deep-digging action, or the use of subsoiling attachments, a single-furrow plough may well constitute a full load for the tractor. On some ploughs the lateral spacing of the bodies is adjustable to give an alternative basic furrow width, although advantage is not always taken of this facility.

Attachment of plough to tractor and preparation for work

PREPARING THE TRACTOR

To prepare the tractor before attaching the plough, the following checks are necessary:

Wheel setting

When the plough is attached to the three-point linkage of the tractor, it is virtually part of the tractor. Because the offside wheels of the tractor—and with reversible ploughs the near side too, on alternate bouts—run in the open furrow trench left on the previous bout, the setting of the tractor wheel-track width determines the width of the furrow slice cut by the front body of the plough (Fig 18).

Consider, for example, a right-hand plough. If the tractor wheels are too narrow, the plough will be positioned too far to the off side (or right). The lateral distance between the front disc coulter and the wall of the open furrow trench will be narrower than that between the front coulter and the second coulter. Conversely, if the tractor wheels are set too wide, the plough will be positioned too far to the near side (or left), thereby increasing the width of the front furrow.

The cross-shaft screw lever, i.e. the front-furrow width control

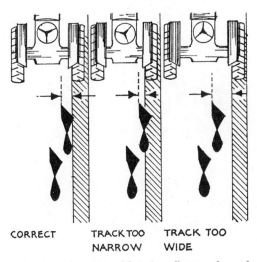

CORRECT   TRACK TOO   TRACK TOO
NARROW   WIDE

18 Tractor wheel-track setting. Note its effect on front furrow width

on the plough, is provided to correct *minor* discrepancies of front-furrow width, but this should not be used as a substitute for correct wheel-track adjustment.

Linkage and check chains

On some tractors, alternative vertical positions are provided for the forward ends of the lower links of the tractor hydraulic linkage system. On others, telescopic lift rods provide alternative lower-link settings. Particulars of these settings are specific to each model of tractor and/or plough, and the maker's recommendations must be followed. Check chains (Fig 19) should allow some latitude for the plough to find its own line of draught, but at the same time they should prevent excessive side-swing of the plough when raised.

19

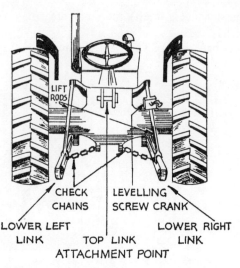

19 Check chains

Tyre pressures

Tyre pressures must be checked before work is started. Reducing the pressure increases tyre deflection and so improves adhesion. In most cases pressures as low as 10 to 12 lb per square inch are permissible. However, further reduction might result in damage to the tyre walls, especially when they are subjected to the weight of the plough in the raised position.

When a mounted reversible plough is to be attached, it is important that the rear tyres have equal pressures, otherwise the level of the plough will be upset as the outfit alternates from left-hand to right-hand work, and repeated correction of the plough level adjustment will be necessary at the commencement of each bout.

PREPARING THE PLOUGH (RIGHT-HAND TYPE)

Provided the plough is in good mechanical condition, its preparation amounts merely to the rotation of the cross-shaft by means of the cranked screw lever so that its cranked ends are in the vertical plane (Fig 8a). This simplifies alignment of the lower links of the tractor when it is backed up to the plough. If the wheel-track of the tractor is correct, this neutral position of the cross-shaft should automatically be its basic position when the outfit is in work, thus providing ample scope for any front furrow width correction necessitated by undulating land or the formation of opening ridges and finishing furrows.

ATTACHMENT PROCEDURE

Fully mounted ploughs

In normal circumstances, manhandling of the plough can be avoided by accurate positioning of the tractor. There are two standard sizes of attachment points in current use—these are classified as *Category* I and *Category* II—the former having a maximum lower hitch pin diameter of 0·870 in, and an upper-pin diameter of 0·750 in, and the latter diameters of 1·115 in and 1·000 in respectively. Conversion devices or alternative attachment points are available to match tractors and implements whose categories differ.

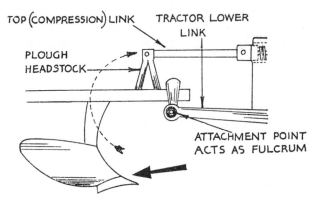

20 Effects of soil resistance on mounted plough
(as tractor moves forward, soil resistance causes plough to rotate about its attached points, causing compression in the top link)

In making the three-point attachment, the following sequence is usually best: (a) lower left link, (b) lower right link, (c) top link—first at the plough headstock and finally at the tractor clevice. It is usually necessary to 'inch' the outfit forward or back slightly in order to line up the front connection point of the top link. With some tractor/plough combinations, however, it is easier to attach the front end of the top link first.

On some ploughs, alternative lower-link attachment points are provided—the higher position for use when plough penetration may be difficult due to hard soil conditions, and the lower position for when tractor wheel-spin may be the problem. The top link is sometimes called the *compression link*, because of the nature of the forces which act through it when the outfit is in work (Fig 20). It is adjustable lengthwise, and on some ploughs alternative attachment points are provided on the headstock (Fig 21).

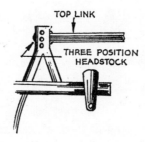

21 Alternative top link attachment points on plough headstock

Semi-mounted ploughs

There is a lack of standardisation of methods of attachment of semi-mounted ploughs, so that generalisations are not very helpful. Briefly, such a plough is attached either by a positive connection to the tractor drawbar, or by connection to the three-point linkage whilst the rear of the plough is supported by castor wheels. Accurate alignment of the plough to the tractor is just as important in this case—indeed the more rigid attachment of the plough to the tractor makes correct alignment imperative.

Remote hydraulic rams are employed on some reversible

ploughs to operate the turnover action. Also, some large right-hand multi-furrow ploughs employ a hydraulic ram to raise the rear end (Fig 15). Where these are used, care should be exercised when coupling the hydraulic hose to prevent dirt getting into the system, and to ensure oil-tight unions.

Operational adjustments

PLOUGHING DEPTH

Ploughing depth may be controlled by a cranked screw lever on the plough designed to raise and lower a land wheel (Fig 22)—an arrangement which ensures consistency of ploughing depth in spite of undulating land and variations in soil texture. In many cases, however, the plough is totally suspended on the tractor

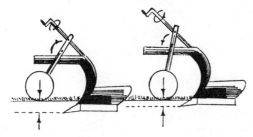

22 Land wheel control gives consistency of ploughing depth: *left* Set deep, *right* Set shallow

hydraulic linkage, and the working depth is regulated through the hydraulic system by a control lever on the tractor.

As the compressive force acting through the top link is proportional to the soil resistance (Fig 20), it is widely used to provide automatic ploughing depth control through the medium of the hydraulic system of the tractor. The stiffer the soil is, the harder the top link is pushed forward, and the stronger is the signal passed to the hydraulic system. In response, the plough is then raised by an amount depending upon the operator's presetting of the hydraulic controls. When soil resistance decreases, the plough

23

is lowered again. It will be seen that selection of the alternative vertical positions for the top link headstock connection (Fig 21) can be made to give a more or less sensitive reaction as required. However, where the soil texture varies in any one field it is usually necessary for the tractor driver to make corrections on the hydraulic control to keep the depth consistent. Sometimes this depth control by plough draught (*draught control*) is incorporated in the lower links of the tractor to achieve greater accuracy of automatic control.

Further refinements in tractor hydraulic systems enable the operator to regulate the amount of implement weight transferred on to the tractor wheels, so that he can select at will either maximum weight transference—on ground causing wheel-slip, but without troublesome undulations—or, alternatively, maximum consistency of ploughing depth where the land is badly undulating but where traction is good (Fig 23). An infinite variation between these extremes is also possible.

23 Soil forces reflected in the top link can be used to achieve automatic depth control on undulating land

One system combining both land-wheel and hydraulic-depth control provides consistency of ploughing depth with controlled weight transference.

The best depth to plough depends on variables such as soil type and condition, and the requirements of the intended crop. Although ploughing depth often determines the worked soil available for the subsequent seedbed, it is sometimes limited by the depth of the top soil which exists on the land to be ploughed. On some land a small proportion of subsoil is from time to time deliberately brought to the surface with the object of weathering and intermixing it with the top soil to increase gradually the depth of the latter.

LEVEL OF PLOUGH

With conventional ploughs, level adjustment is made on the tractor right-hand linkage lift rod (Fig 19). Rotation of a cranked lever alters the length of the rod, and so raises or lowers the right-hand side of the plough. For normal ploughing, the levelling screw should be adjusted so that the vertical portion of the plough beams, when viewed from the rear, is perpendicular to the ground surface (Fig 24a). This does not apply when marking out the headland and

24 Rear view of tractor and plough. *a* For normal ploughing vertical beams should be perpendicular to ground surface. *b* For marking out headland or opening out a ridge beams should be tilted until only one share is ploughing

when making the first run of a ridge, since for these operations the plough is intentionally tilted (Fig 24b).

With most reversible ploughs, the level adjustment is located on the plough-head assembly and the right-hand lift rod adjustment should not be disturbed from the prescribed basic setting.

FRONT FURROW WIDTH

Adjustment of front furrow width on most ploughs is by a control that rotates the cross-shaft (Fig 8). This has the effect of angling the plough laterally, either leftwards directing the shares towards the unploughed land, or rightwards away from the unploughed land (using right-hand bodies). The adjustment is meant to correct minor variations in front furrow width which may occur when side gradients are encountered.

If there is a general discrepancy of more than 2 in in front furrow width when the outfit is on level land with the cross-shaft in mid-position, the tractor-wheel setting is probably incorrect. An attempt to rectify more serious front furrow width discrepancy by means of the cross-shaft adjustment will result in the plough *crabbing*, i.e. adopting an angle out of line with the tractor. This causes undue pressure and wear on either the mouldboards or the landside plates, depending upon the direction of the maladjustment. In either case, the quality of the work would be poor and the draught would be increased. Another ill effect may be felt in the steering of the tractor due to the *side draught*, i.e. lateral forces resulting from the misalignment of the plough to the tractor. Most tractor and plough manufacturers supply information on the correct matching of tractor to plough, and this should be referred to.

Another factor which can influence tractor/plough alignment is the provision on some ploughs of a sliding cross-shaft. Whilst this should normally be set to position the plough headstock in line with the centre of the tractor, it is permissible to offset the plough to get close to a headland boundary when expedient.

REVERSAL CONTROL (REVERSIBLE PLOUGHS)

The reversal control alternates the bodies of the reversible plough at the end of each bout. It may be manual, or powered mechanically or hydraulically. The manual type of turnover mechanism is usually adequate, although sometimes the adherence of soil to the mouldboards upsets the balance of the bodies on which the mechanism relies. With some mechanical lifts, the changeover is achieved automatically as the plough is lifted out of work at the bout ends. The hydraulic type employs a remote ram actuated by the tractor hydraulic system.

TOP LINK ADJUSTMENT

Although in some cases the top link is provided with a means of adjustment without the aid of tools, this setting adjustment for conditions generally should always be regarded as a critical basic setting, and not one to be constantly interfered with in any one field. Its exact setting must be made with the plough in work and at the intended ploughing depth. Correct top link length can be checked by observing the landside plate of the rear body. The plate should be lightly contacting the bottom of the furrow trench (Fig 25a). If its rear end is digging into the bottom of the furrow,

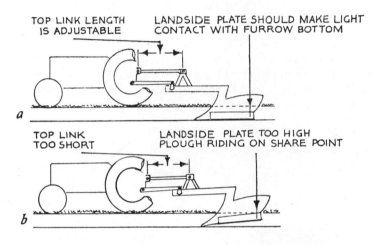

25 Adjustment of top link length: *a* Correct, *b* Too short

the top link is too long; if, on the other hand, the landside plate is riding clear of the furrow bottom, the link is too short (Fig 25b). Excessive pitch caused by a short top link will accelerate wear of the shares, and produce a rough furrow bottom. Although the plough pitch may be increased slightly to improve penetration in hard conditions, it should only be done when expedient. Normal setting should be reverted to as soon as possible.

ADJUSTMENT OF DISC COULTERS

The function of the coulters has already been outlined. For satisfactory operation, their setting in relation to their respective shares must be varied to suit the working conditions. It is not possible to lay down hard and fast rules for the precise clearances

27

which should be provided between the disc and the share, but the following basic settings provide a starting point from which fine-settings can be made. On most ploughs the two usual forms of adjustment are vertical and lateral, while on some ploughs longitudinal and tilting settings can also be made.

Vertical clearance

A basic clearance of about 2 in between the share and the lower edge of the disc, at their nearest points, should be sufficient when ploughing at a depth of 5 in to 8 in in normal soil conditions (Fig 26). The coulters should be *raised* to increase this vertical clearance in the following circumstances:

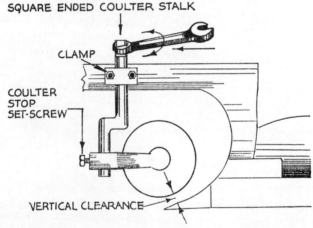

SQUARE ENDED COULTER STALK

CLAMP

COULTER
STOP
SET-SCREW

VERTICAL CLEARANCE

26 Vertical coulter clearance (side elevation)

1. Deep ploughing—to prevent the coulter arms fouling the ground surface, and to prevent the coulter bearings carrying excess plough weight.
2. Long straw, loose trash or manure on the surface—to prevent choking of the plough.
3. Hard, dry and stony soil conditions—to prevent the plough from riding on the coulters, as this impedes plough penetration and unduly loads the coulter bearings.

28

The coulters should be *lowered* from their basic setting in the following circumstances:

1. Shallow ploughing—to ensure that the furrow edges are cleanly cut.

2. Marking out field, forming ridges and finishes. (Normally only the rear coulter setting need be adjusted for these operations.)

3. Ploughing grassland or clover—to ensure clean cutting of the fibrous roots.

Lateral clearance

For a cleanly cut furrow wall, a certain amount of side clearance is necessary between the cut made by the coulter and the line taken by the landside face of the share (Fig 27). This is generally

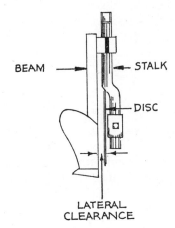

27 Lateral coulter clearance (front elevation)

obtained by rotating the cranked stalk which supports the coulter arm (Fig 26). A basic setting in normal conditions is about ¾ in, though this, too, may be varied considerably according to the nature of the work on hand. The following conditions require a slightly increased clearance:

1. Hard, stony land.

2. Wet, sticky land.

3. Loose soil (such as may be encountered after a root crop or when cross-ploughing).

The lateral setting of the coulters is commonly adjusted to maintain uniformity of furrow width. But, if a discrepancy exists with all the coulters set equidistant from their respective shares, the cause of the trouble lies elsewhere. Possible mechanical faults are discussed under *Maintenance*.

On some reversible ploughs, the coulters are carried on double-cranked stalks so that a single adjustment affects the coulters of two opposing bodies simultaneously. This ensures uniformity.

Most coulter stalks have an adjustable stop collar designed to restrict the side swing of the coulter arm and disc (Fig 26). The stop should be set so that the disc has some freedom to swing in either direction out of its line of travel, particularly when working in hard, stony land. If the stop collar is inadvertently left slack, there is a danger of the disc swinging right round when the plough is raised, and being damaged when the plough is subsequently lowered into work.

Longitudinal adjustment

On those ploughs provided with forward and backward coulter adjustment, the normal setting should be such that the centre of the disc is approximately in line with the point of the share (Fig 26). It is an advantage to move the coulter back when ploughing hard, dry land, as this allows the share to penetrate the hard surface crust of the soil before the coulter takes any of the plough's weight.

Tilt adjustment

When tilting the disc, it is usual to rotate it 10° to 15° in a clockwise direction (viewed from the rear). This has the effect of under-cutting the furrow wall and decreasing the angle of the exposed corner of the furrow slice (Fig 28). The advantages are as follows:

1. It improves coverage of trash—although this technique is not always a substitute for the use of skimmers as is sometimes supposed.

2. It is an asset where the ploughing has to be worked down into a seed bed immediately.

3. It produces the type of work which settles flat with subsequent rolling, so reducing air cavities between the furrow slices and the subsoil—a useful technique for late spring or summer ploughing when immediate sowing is to follow. For the best results, the angle of tilt should be about 15° (although less in hard soil conditions).

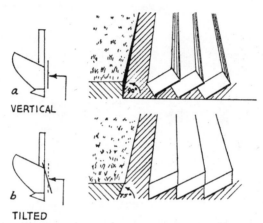

28 Effects of tilting coulters. Note good overlap of furrows (*a*), high crest of furrows and minimal overlap (*b*)

Objections to the use of tilted coulters as standard practice—two of which are of a mechanical nature—are these:

1. It is unsuitable as a winter ploughing technique, because the furrow slices tend to sink into the furrows and lie flat.

2. Undercutting of the furrow wall increases the abrasion of the shin or the leading edge of the mouldboard on each body.

3. The angle of the discs imposes undesirable end thrust on their spindles, particularly in hard soil conditions.

ADJUSTMENT OF KNIFE COULTERS

Knife coulters, like disc coulters, should be offset a little from the

31

line taken by the landside face of the share in order to achieve a clean furrow wall. A vertical clearance of 1 in to 2 in is usually satisfactory, and in the basic longitudinal setting, the leading edge of the coulter at its lower end is approximately in line with the share point.

ADJUSTMENT OF SKIM COULTERS (SKIMMERS)

Although skimmers can be troublesome under difficult ploughing conditions, they should be used more than they are at present (Fig 29). Despite their usefulness, many ploughmen remove them rather than take the trouble to adjust them.

29 Action of skim coulter: surface vegetation is skimmed, deflected into furrow trench and subsequently buried by furrow slice

The method of adjustment depends on the type of fitting. On some ploughs fitted with disc coulters, the skimmer attachment is secured to the coulter arm. On all other ploughs, the skimmer is carried on an independent stalk secured to the plough beam (Fig 30). The latter fitting is more satisfactory as it gives greater scope for longitudinal adjustment, enabling the skimming to be taken off ahead of the coulter and before the furrow slice starts to rise on the breast of the plough. When the disc coulter and skimmer are both on the same arm, choking occurs in trashy conditions because stones and crop residues are unable to filter between the disc and the skimmer.

When setting skimmers, bear in mind the following points:

Depth

In setting skimmers, depth is critical. Variations of ploughing depth while ploughing a field will automatically vary the depth of the skimming. Where the surface of the land is reasonably clean and level, the depth of the skimmer should not exceed 2½ in. When ploughing in lands, different settings will be necessary for forming the ridges, making the finishes and for the routine work.

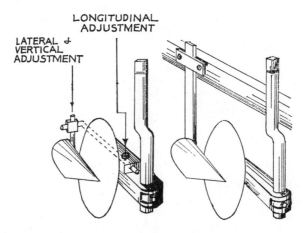

30 Alternative skim coulter fitments

Longitudinal adjustment

The skimmer should be set well forward to remove the strip of surface growth before the furrow slice starts to rise. This produces consistent skimming, and correct deflection of the trash into the furrow bottom (Figs 29 and 31)

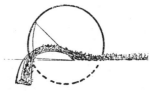

31 Action of skim coulter

Lateral adjustment

When the skimmer is integrated with the disc coulter, its point

33

should be almost touching the side of the disc whilst its top edge should have about ⅜ in clearance (Fig 32). Sometimes, the angle of

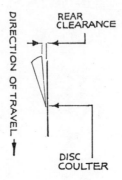

32 Plan view of skim coulter in relation to disc

the skimmer blade can be adjusted to provide greater deflection of the skimming. When used with knife coulters, the skimmers may be positioned to clear any trash in the path of the coulter.

MOULDBOARD ADJUSTMENT

This facility is not provided on all makes of plough. Even where it is provided it is commonly misused, often in an effort to correct faults caused by more superficial maladjustments. Mouldboard adjustment is valuable as a maintenance provision to obtain uniformity of soil deposition.

On some ploughs the adjustment is simple and may be effectively used as an operational setting. Pushing the ends of the mouldboards out produces a more pulverising effect on the soil, and drawing them in helps to prevent furrows falling back when ploughing heavy soil or grassland, or when turning furrows up a gradient. The explanation of this is that if the furrow slice is kept as straight as possible, the twisting action of the mouldboard is more effective. Also the 'narrower' the mouldboards are set, the flatter the furrow slices will be (Figs 33 and 34).

34

Altering the angle of the mouldboard varies the width of the furrow trench in which the wheels of the tractor run. So extra-wide tractor tyres can be prevented from scraping the last furrow slice by moving all mouldboards out a few degrees. On wet clay

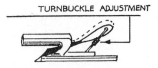

TURNBUCKLE ADJUSTMENT

33 Mouldboard adjustment

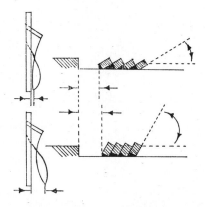

34 Effects of mouldboard angle on deposition of furrow slices and width of trench

land, however, mouldboards should not be moved out too far because the increased pressure of the mouldboards will have a smearing effect on the furrow slices. Excessive angle also causes additional friction which increases draught and accelerates mould-board wear especially on abrasive soils.

BASIC FURROW WIDTH ADJUSTMENT

Reference has been made to factors affecting the width of furrow produced by the front body of the plough. However, the basic furrow width depends on the relative lateral spacing of the bodies of the plough. On some ploughs, the beams and their supporting struts are so arranged that no adjustment is possible; on others, alternative spacings are easily obtained. Where such a provision is made, full advantage should be taken of it. In light working conditions, increasing the furrow width increases the working capacity of the plough. In difficult conditions, reducing the furrow width may prevent wheel-spin, and the narrower furrows on wet land may also prove an advantage in subsequent cultivations.

The various operational adjustments of ploughs are discussed further under *Operational checks*.

Special attachments

SPECIAL DISC COULTERS

The introduction of combine harvesters has presented farmers with the problem of ploughing in long stubble and loose straw. Several devices have been marketed to simplify this work, but probably the most effective plough adaptations are the wavy-edged and scalloped-edged disc coulters which are rotated more positively by ground contact than are the standard plain discs (Fig 35).

35 Wavy-edged and scalloped disc coulters for trashy conditions

MOULDBOARD TAIL-PIECES

Mouldboard tail-pieces are extensively used on modern ploughs. They prevent furrows falling back when the plough is on stiff land or working across a gradient (Fig 36). However, on heavy, wet land, they do sometimes tend to smear the furrows, and then they are probably best removed.

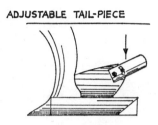

36 Mouldboard tail-piece

One or two types of extension are available which incorporate tines designed to cut and shatter the furrow slices where this is desirable.

CHAINS

Though not strictly special attachments, chains are very effective when ploughing in long clover, mustard or similar crops, which have a large volume of foliage to bury. The chains are attached to the plough beams well in front of each body. Their length should be such that they drape over the furrow slices as the latter are turned, with their ends gripped by the inverted portion of the furrow slices. The chains thus drag the foliage back and downward and assist its burial.

SUBSOILING ATTACHMENTS

Subsoiling attachments are usually tines, fixed on the plough to the rear of each share. A narrow, pointed share at the bottom of each tine breaks up the hard, top layer of subsoil which impedes drainage, aeration and the penetration of plant roots (Fig 37).

37

Field procedure

Timing is an important factor in most seasonal operations on the farm, and although ploughing is perhaps not as exacting as many of them, its effectiveness can be greatly influenced by timing. The type and condition of the soil, the intended crop and the

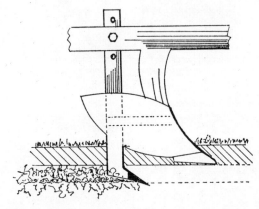

37 Subsoiling attachment fitted to plough. Tine breaks up subsoil behind and beneath ploughshare

nature and condition of any vegetation to be buried have to be considered. Other factors permitting, it is better to plough in growing vegetation than ungerminated weed seeds.

Selection of the plough for any particular operation should also be made with these factors in mind.

PLOUGHING SYSTEMS

There are a number of systems by which a field may be ploughed. In selecting one, the type of plough, the size and shape of the field, any undulations of the land, any drainage problems, and any special needs of the seedbed and after-cultivations, must all be considered.

Whatever system is employed, however, it is important to adopt a definite plan of procedure. The various methods are generally referred to as *in lands* (ridge and furrow), *one-way*, *square*, and *round and round*.

In lands

Ploughing in lands has been the almost universal method for many years. Because most ploughs have been capable of throwing the soil in only one direction, continuity of work could only be achieved by ploughing in plots or *lands*. The general sequence for ploughing in lands is as follows:

Marking out the headland

The headland marking line is normally a single shallow furrow ploughed with the rear body of the plough in a clockwise direction around the field parallel to the boundary (Fig 38). This margin is

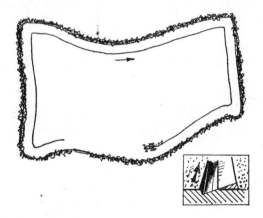

38 Headland marker furrow and shallow skim furrow produced (*inset*)

necessary for turning the outfit at the ends of the straight runs. It should be wide enough to enable the outfit to get properly into line with the furrow when entering each bout, and to eliminate the need for time-wasting shunting which can sometimes be precarious near banks and ditches. An adequate headland width also avoids unnecessary 'independent' braking. Eight yards is satisfactory in most cases. The marking furrow serves as a guide to the operator when lowering and raising the plough into and out of work.

Normally, the marking furrow should be as shallow as possible. When the land is dry and hard, however, and entry of the plough difficult, a deeper headland furrow should be ploughed in the opposite direction (anti-clockwise). This will present a furrow wall to the plough-share points on entry and thus simplify their penetration under the hard surface.

The novice may find it difficult to gauge the headland width, particularly where the field boundary is irregular. In such circumstances, it is probably better for him to pace out the width at frequent intervals, especially when grassland has to be ploughed and clean bout ends are imperative. Alternatively, the headland can be marked out with a second man using a length of twine as a gauge. The margin of a stubble field can often be accurately gauged by following a stubble drill of the previous crop. With very sharp corners, it is a good idea to plough a few short furrows at the apex to give a reasonable radius for subsequent cornering when ploughing the headland later.

Setting up the opening ridges

Setting up the opening ridges is an important operation, and one requiring both forethought and skill. In a field that is more or less rectangular, it is usually most economical to form the ridges lengthwise. However, if the land is on a hillside, or if the hedges are very irregular, it may be better to plough the field parallel to one of the shorter sides (Fig 39).

The spacing of the ridges depends on the size and shape of the field. Ridges set too far apart result in too much traversing and compacting of the headland, especially where a field boundary lies oblique to the general direction of the ploughing. Setting the ridges too close together, on the other hand, is time-consuming, and disturbs unnecessarily the levelness of the field surface, especially if the ridges and finishes are badly made.

A land width of 12 yd for each body on the plough is a convenient basis on which to work. A different width may be necessary if there are obstructions in the field, or if the side-land boundary does not lie along the general line of ploughing, causing the bout length to vary.

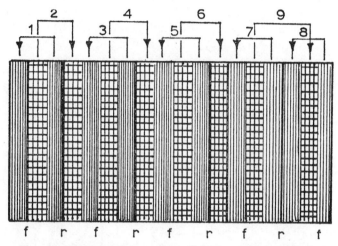

39 Ploughing 'in lands'—sequence of the 'casting' and 'gathering' phases when ploughing

Two types of *ridge* are commonly used—the *double-split* and the *single-split*. The former is preferable if it can be formed successfully. The latter is the best alternative method of opening out ploughing, and is often used in the ploughing of grassland.

Double-split ridge

If a double-split ridge is properly formed, no land is left unploughed beneath it. When the ridge spacing has been decided upon, and markers placed to ensure a straight opening furrow, the plough should be set to cut a shallow furrow with the rear body only. This is done by raising the front and any intermediate bodies with the levelling screw until they are just clear of the ground, as for the headland-marking furrow. If the front bodies are lifted too high, a very narrow rear furrow trench results. The rear coulter should now be lowered until its bottom edge is level with the *bottom* of the share. This ensures that a clean furrow wall is produced. Work can now proceed as follows:

1. The outfit should be positioned on the headland mark and correctly aligned to make the first run of the ridge (Fig 40a). The plough should be lowered to the minimum depth which gives a continuous cut—a depth of 2 in to 3 in will suffice in good ploughing conditions where the surface is clean, but in rough or trashy

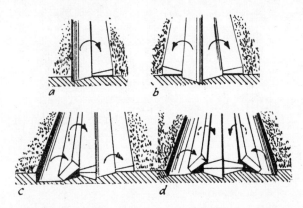

40 Formation of double split ridge. Note how central furrow slices are cut and then replaced. *a* First run: shallow skim furrow. *b* Second run: slightly deeper. *c* Third run: both bodies ploughing. *d* Fourth and final run: both bodies ploughing

conditions it may be necessary to plough considerably deeper to ensure a continuous furrow. After a short run, the outfit should be stopped and a check made on the work being produced.

2. At the end of the first run, the outfit should be turned round and aligned to make a return run with the tractor wheels straddling the furrow trench just opened up. The front bodies should be kept raised, but the rear body should be set about 1 in deeper to help stabilise the plough and prevent side-swing. The tractor should be lined up so that the rear body of the plough cuts a second furrow trench immediately alongside the first (Fig 40b).

3. For the third bout, the plough should be levelled so that all bodies are brought into work, and the rear coulter should be raised to a higher setting. The outfit should be driven with the off-side wheels of the tractor travelling in the left-hand furrow trench, so that the front body of the plough replaces the single shallow furrow slice in its path together with a new shallow furrow cut from beneath it. *Ground speed must be kept low*, and care must be taken on this run to ensure that the front 'sandwich' furrow slice occupies only *half* of the open double furrow trench (Fig 40c).

4. On the fourth and final bout, the setting of the plough should remain as for the one just completed. As before, the front body replaces the remaining shallow skim furrow slice together with a new one cut from beneath it. During this operation the front 'sandwich' furrow must be kept within the remaining open portion of the double trench (Fig 40d).

Although the ridge is now complete, the maximum general ploughing depth will not yet have been reached but graduation to this will take place over the next two bouts of routine ploughing, after which both the coulter and skimmer settings should be checked and modified as necessary.

Single-split ridge

With a single-split ridge, a narrow strip of land must inevitably remain unploughed, but any grass or other vegetation present should be completely sealed by the covering furrow slices. The procedure is as follows:

1. On the first run, as with the double-split ridge, a single shallow furrow is ploughed—again at low ground speed (Fig 41a).

2. On the return run, however, the plough is levelled and all bodies are brought into action. The outfit is positioned so that the front body places its furrow slice to cover half the initial shallow furrow slice (Fig 41b). If the tractor is driven with its off-side wheels running on the initial furrow slice, this helps to produce a flat and tightly compacted ridge.

3. The third and final bout usually requires slight lateral adjustment of the plough relative to the tractor so that the front furrow

slice is placed snugly edge to edge with the front furrow slice of the previous bout (Fig 41c).

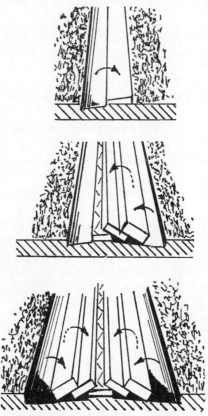

41 Formation of single-split ridge. Note tractor wheel tread mark on initial furrow slice to aid compaction and sealing of unploughed strip

Routine ploughing

Reference has been made to factors which normally influence the choice of ridge spacing. This in turn will determine the general sequence adopted for the routine work. Fig 39 illustrates a system commonly used where the ridges are set up parallel to a straight field boundary. The indicated phases of *casting* and *gathering* avoid awkward turning of the outfit, and excessive traversing of the headland.

The width of the first land is three-quarters that of the remaining lands, and the outfit *casts* furrows until a quarter of a land width has been ploughed against each of the first two ridges. (*Casting* means that the outfit turns to the left at the end of each bout, so that the furrows are cast outwards, away from the unploughed land around which the outfit is working.) At the end of this phase, a quarter of a land width will remain unploughed between the first two ridges. The outfit now proceeds to *gather* furrows (i.e. turn to the right at the end of each bout) around the second ridge until the remaining portion of the first land has been ploughed. The second land will then be reduced to three-quarters of a full land width, and the whole procedure begins again (casting until one quarter of a land has been ploughed on either side and then gathering until the land has been completed).

When the field has been reduced to two land widths, the outfit should cast in each of the lands until a uniform width of unploughed land remains in each. The work can then be completed by gathering around the dividing ridge (note phases 7, 8 and 9, Fig 39). Various other systems of ploughing in lands are possible, but all the alternatives cannot be detailed here.

Whilst carrying out the routine ploughing, ground speed should be carefully regulated, as this influences the type and quality of work considerably. Also the plough should be lowered and raised consistently on the headland mark.

Finishes

The *finish* is the furrow that results from the matching up of one ploughed section with the next. In order to achieve a good clean finish, with no land left unploughed, no land double ploughed and no unnecessary wheel marks on the ploughing, the opposing furrows must be kept parallel during the routine work. This is why accuracy in marking out the field, setting up the ridges and ploughing out the lands, is so important. In any event, it is advisable to check the width and parallelism of the unploughed strip in good time for corrections to be made before the final few bouts are reached. By taking a slightly wider. or narrower front furrow where necessary, considerable discrepancies can be rectified.

On the penultimate bout, the plough should leave a strip of unploughed land behind it of a width equal to one furrow *less* than the capacity of the plough. Thus, a plough having two furrows of 11 in width should leave a strip of 11 in to be ploughed on the finishing bout (Fig 42a). This is necessary because, on the last bout, the plough has no rear furrow wall to counter the side-thrust. During this final operation the front body or bodies of the plough will be in action, while the rear body will be running light

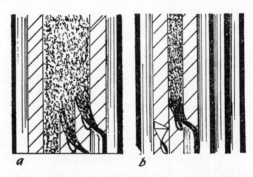

42 The *finishing* furrow: *a* Second last bout with two furrow plough—leaving one furrow unploughed; *b* Final bout—front body only ploughing, rear running light in open furrow trench

—apart from skimming the empty furrow trench to produce a little loose soil so that the furrow is not left devoid of tilth (Fig 42b). Also, if any land remains unploughed after the forward bodies have passed, the rear body will cut and lay this over. As the final few bouts are approached, it is important to reduce the ploughing depth gradually. This avoids deep furrows across the field which can damage harvesting machinery, and present serious problems where after-cultivations have to be performed on row crops.

In the ploughing of grassland or clover ley especially, the finish calls for a high degree of skill and accuracy.

Ploughing the headland

When the routine ploughing has been completed, work can start on the headland. This is often done hurriedly and without the care it justifies. Where the marking furrow has been accurately drawn and carefully adhered to, the operation is straightforward, and light running is reduced to a minimum. The headland may be ploughed in either a clockwise or an anti-clockwise direction (Fig 43). It is common practice to reverse the direction each year,

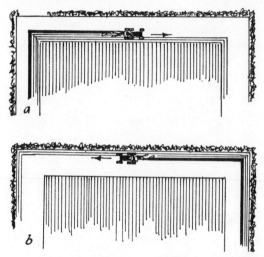

43 Alternate methods of ploughing headland: *a* Clockwise direction—*gathering*, *b* Anti-clockwise—*casting*

to prevent a ridge of soil building up, either at the field boundary or at the ends of the straight work. Ploughing the corners is simplified with mounted ploughs, and by offsetting such ploughs on the cross-shaft it is possible to plough close to the field boundary.

Where drainage problems are likely to arise, the open finishing furrow trenches should be continued across the headland at the lower end of the field to the ditch, after the headland has been ploughed.

C

ONE-WAY PLOUGHING

One-way ploughing is carried out with the reversible plough. It avoids the time-consuming jobs of marking out lands, setting up ridges and ploughing finishing furrows.

It is possible to avoid sidelands by ploughing against one field boundary and progressing across the width of the field to the other boundary, leaving headlands at the bout ends only. But it is preferable to mark out a margin all round the field rather than to drive on the ploughing, as invariably occurs when operators 'manage' with end headlands only.

One-way ploughing also eliminates the need for straightness—a feature which has been the source of so much pride and ridicule on the land for many years (Fig 44). However, this advantage of

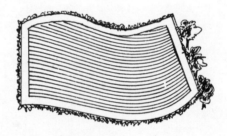

44 Field ploughing with reversible plough. Bouts have followed boundary at bottom of Figure

the reversible plough is no excuse for careless or untidy workmanship. In fact, the correct setting of a reversible plough calls for considerable skill, and constant alertness is necessary as work proceeds.

If a field has a very irregular shape, it may be preferable to plough it in two or more sections.

SQUARE PLOUGHING

Square ploughing is often adopted for large, conveniently shaped fields, where it shows a considerable saving in time and fuel over ploughing in lands. Provided the ploughing at the corners is neatly matched, it produces a much more level field surface than ploughing in lands.

First, a small replica of the field's shape is marked out in the centre. The importance of accuracy in this operation cannot be over-stressed. One satisfactory method is that sometimes employed for the marking out of headlands—a length of twine used as a gauge and a second person to walk round the field boundary. The procedure may have to be repeated a number of times, using the previous shallow marker furrow as the perimeter at each stage until the replica is reduced to a convenient size.

This small central area is then ploughed by setting up a ridge and gathering around it (Fig 45). The outfit then ploughs along

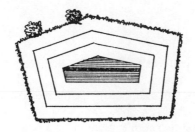

45 Square ploughing. Marker furrows and ploughed central replica

each side of the area in a clockwise direction, negotiating the corners either by *looping* or *shunting*. The accuracy of the work in progress can be checked by comparing the ploughed area with the marker furrows made during the marking-out process.

ROUND AND ROUND PLOUGHING

In round and round ploughing, as in square ploughing, the routine work follows the field boundaries. The important difference, however, is that in this case the work is done in an anti-clockwise direction. The outfit starts by ploughing around the field close to the hedge or fence, and simply continues until the centre of the field is reached. The corners may be negotiated in one of two ways—either by picking up the plough and leaving diagonal strips to be ploughed as the final operation (Fig 46a), or by ploughing right out on each side and reversing at the corners to align the outfit for the next side (Fig 46b).

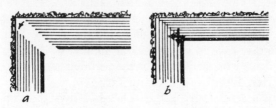

46 Alternative methods of cornering in *round and round* ploughing: *a* Unploughed diagonal strip, *b* Shunting method of cornering

OPERATIONAL CHECKS

Of all the tractor driver's many jobs, few are as fascinating and gratifying as ploughing, given reasonable working conditions, a good set of equipment and, above all, the necessary skill to set the plough. Conversely, few jobs can be more exasperating to a novice struggling to set his plough by hit and miss methods. Even ploughmen with years of experience sometimes resort to random experimenting when they encounter snags, rather than adopt a systematic sequence of checks to isolate the cause or causes of their problems.

A plough which performs quite satisfactorily in stubble or loose soil may produce poor quality work when operating in grass land or heavy clay, where the firmer furrow slices reflect both the plough's setting and its mechanical condition. Considerable skill is needed to distinguish between the subtle effects of maladjustment of the plough's components and those caused by distortion of its structure.

The following schedule of checks provides a suitable procedure to adopt when a plough is performing unsatisfactorily. First examine the plough for obvious faults, such as excessively worn or broken shares or even complete loss of a share, badly disturbed coulter or skimmer settings, soil sticking to mouldboards, or the possibility of obstruction between a coulter and share.

Ploughing depth/width ratio

When a slow-turning mouldboard is used in firm soil conditions, and where good packing of the furrows is required, the depth should be limited to about two-thirds of the furrow width in order to produce the rectangular furrow slice section necessary for its correct deposition. Fig 47 illustrates the effects of alternative ploughing depth/width ratios. If the ploughing depth is excessive,

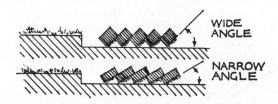

47 Effect of ploughing depth on deposition of furrow-slices: *top* deep ploughing, *bottom* shallow ploughing. Note the respective angles

the furrow slices tend to stand on edge or even fall back. Sometimes, however, the problem can be reduced by using slightly narrower shares, so that the wing of the share leaves a narrow strip of soil uncut to act as a hinge to the furrow slice as it is inverted. The narrow strip is ultimately severed from the subsoil by the turning action of the mouldboard.

Levelness of plough

Levelness can readily be checked by standing behind the outfit with the latter in work, and observing the vertical portions of the beams of the plough. These should be perpendicular to the ground surface, even though the tractor will inevitably be tilted with two of its wheels in the open furrow trench (Fig 24). If necessary, correction should be made by modifying either the tractor right-hand lift-rod adjustment, or the plough-head adjustment (depending on whether the plough is of the right-hand type or reversible type respectively).

Pitch of plough

The pitch is influenced by the top link length. To check the setting, inspect the landside plate of the rear body. Its heel should be just contacting the furrow bottom without gouging a groove in it (Fig 25). This setting is affected by variations in basic ploughing depths, and should therefore be checked for each particular ploughing operation. The pitch may be increased when the soil is hard and dry. In this case the top link should be shortened slightly. Excessive pitch, however, accelerates share wear.

Individual body pitch adjustment is also possible on some ploughs, but this is a maintenance rather than an operational adjustment (Fig 6).

Front furrow width

A precise check of front furrow width can be made by firs ensuring that all coulters are set equidistant laterally from their respective shares, and then measuring the distance between the front coulter and the existing furrow wall, i.e. the side of the open furrow in which the tractor wheels are positioned. This measurement should be identical to those taken from equivalent points on the second and any subsequent bodies of the plough. It is permissible to correct a front furrow width discrepancy of up to 2 in by adjusting the plough cross-shaft. A discrepancy in excess of 2 in would indicate that one of the following two settings is incorrect:

1. Wheel track of tractor. This should have been checked when the plough was attached (Fig 18).
2. The position of the cross-shaft on the beams. It is usually described in this way, but it is probably more correct to refer to the position of the beams on the cross-shaft. If the setting is incorrect, the headstock of the plough will be out of line with the top-link clevice at the rear of the tractor. The setting can be

checked by sighting the top link from the rear, when it should be in line with the centre-line of the tractor.

If, after these checks, the furrows produced are still not of even width, suspect lateral distortion of one or more of the beams. Checks for this kind of damage are described under *Maintenance*.

Reversible ploughs have various provisions for adjustment of front furrow width, but they all consist basically of lateral or angular movement of the head of the plough. Linch-pins and/or set screws are two of the many alternative types of adjustment commonly used. This setting is perhaps even more critical on the reversible plough than on the right-hand type, and so operators should acquaint themselves with the precise instructions issued by the plough manufacturer.

Coulter settings

The basic adjustments of the coulters have already been discussed together with problems likely to demand modifications. Unsuitable settings may be recognised by the following effects:

1. Poor entry of the plough and inconsistent depth of work in hard soil conditions suggests that the coulters are too low and/or too far forward.

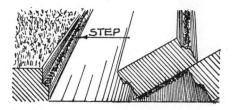

48 Stepped furrow wall owing to *wide* coulter setting

2. Choking of the plough with trash, straw or farmyard manure also suggests too low a coulter setting.

3. Stepped furrow walls indicate that the coulters are set too wide, i.e. there is excessive lateral clearance (Fig 48).

53

4. Torn and irregular furrow walls suggest that the coulters are set either too narrow, i.e. with insufficient lateral clearance from the share, or too high (Fig 49).

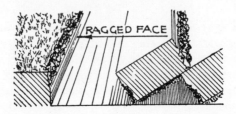

49 Torn and shattered furrow wall owing to *narrow* coulter setting

The two last mentioned faults are fairly obvious when they occur on the last body of the plough, but often they continue unnoticed on the front and intermediate bodies. Operators should keep a constant watch on the work produced by *all* bodies of the plough since coulter settings can easily be disturbed by roots and stones in the soil. Such maladjustments not only detract from the quality of the work, but may also cause excessive wear on the soil-working parts of the plough.

Skimmer settings

The most common error in the setting of skimmers is excessive depth. If too heavy a skimming is removed, it tends to obstruct the furrow slice as it is laid against the preceding furrow. On the other hand, too shallow a setting sometimes results in the accumulation of trash on the skimmer, leading to eventual choking of the plough. If the general ploughing depth is altered substantially, the skimmers should be re-set.

Mouldboards

If the mouldboards fail to place their respective furrow slices uniformly against the preceding ones, a correction must be made (Fig 34). However, first check that the trouble is not being caused

54

by a disturbed coulter setting or by soil sticking to the mouldboard face. The use of mouldboard tail-pieces is not always an
advantage, and a close check should be kept on their effects on the
work.

General uniformity of work

When firm soil is ploughed in reasonably dry conditions, it
should be possible to produce level work with all the furrows of
uniform size, shape and angle of repose; but in soft and wet
conditions, the compacting effect of the tractor wheels on the unploughed land will inevitably be reflected in the appearance of the
furrows which follow the wheel track. This could mislead the
novice who is enthusiastic about his work. Unfortunately, little
can be done to eliminate the effect of compaction.

Maintenance

DAILY ATTENTION

There are few moving parts to a plough, so its lubrication requirements are small. Indeed, on some ploughs lubrication is limited
to the operating controls. On those ploughs which have land
wheels and disc coulters, the bearings of these usually require
daily greasing. In very hard ploughing conditions, the disc
coulters should be greased twice a day, since under such conditions
they are subjected to considerable load.

Before starting a day's work, the shares should be checked for
condition, and the plough inspected generally. Rust preventive
should be removed.

At the end of each day's work, all the soil-engaging surfaces of
the plough should be cleaned and coated with oil or rust preventive. This is particularly important on farms having heavy, sticky
soil.

END-OF-SEASON ATTENTION

At the end of a season's ploughing it is as well to check the plough over, so that any replacements can be obtained in good time for the next period of service. All soil-engaging parts should be inspected. If the wear on the shares and mouldboards is moderate, it may be well worth having worn areas 'hard-surfaced'. If these parts are allowed to wear excessively, reclamation may not be possible. The coulter bearings should be checked for wear— it is not difficult to dismantle the assembly so that the spindle and bush may be cleaned, thoroughly examined and reassembled. When this is done, the condition of the dust felts or seals must be checked, and they should be renewed if necessary. If the discs are worn so that their diameter is reduced by 2 in or more, they should be replaced. Small discs reduce the effective ploughing depth, and tend to cause choking in trashy conditions.

Checking for lateral distortion

A simple method of checking for lateral distortion is to suspend a plumb-bob from the face of the beam above each plough share in turn, and note the distance between the bob and the landside face of the share (Fig 50). The measurement for each body should be identical. For accuracy, new shares should be fitted to all bodies before making this check.

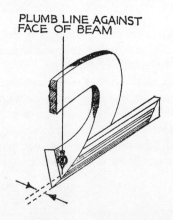

50 Checking for lateral distortion. Clearance between arrows should be the same on all bodies

An alternative method, where a concrete floor is available, is to project chalk lines out from the landside face of each body in turn, and measure the distance between the chalk lines at their extremities. The projection of the chalk lines shows up any discrepancy between the bodies. For this check, both new shares and new landside plates should be fitted. With all bodies lowered to the ground, a straight edge about 8 ft in length should be placed against the side of the share and landside plate of each body in turn (Fig 51). The end of the straight edge should be flush with

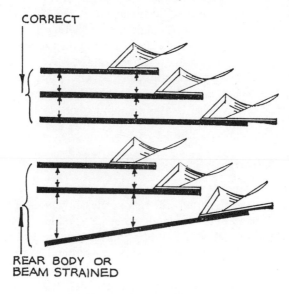

CORRECT

REAR BODY OR
BEAM STRAINED

51 Straight edges used to check distortion by exaggeration

the rear end of the landside plate, on the front and any intermediate bodies. On the rear body, however, because the landside plate is longer than that of the others and sometimes slightly angled, the straight edge is not placed flush with its end. Instead, it should be positioned the same distance behind the point of the share as on the other bodies. The chalk lines are drawn on the floor against the straight edge, so indicating the attitude of each body in turn. If any line is not parallel with the others, distortion is indicated.

Correction of minor discrepancies can sometimes be made on the farm by packing the frog where it is secured to the beam, but this is usually a job best left to the agricultural engineer.

Checking for vertical distortion

After fitting new shares to all bodies, check the under-beam clearance, i.e. the distance between the point of the share and the underside of the horizontal beam above it (Fig 6). All bodies should have the same measurement. When making this check, it is important to hold the rule at a right angle to the horizontal beam. On some ploughs, minor discrepancies can be corrected by means of individual body-pitch adjustments. On some two-furrow ploughs, one body only has adjustment so that it can be matched to the static one. Where body-pitch adjustment is not provided, correction is again best left to the agricultural engineer.

Replacement parts

The main wearing parts of the plough are, of course, those which engage with the soil, namely the shares, the coulters, the skimmers, the mouldboard and the landside plates. Whilst it is always advisable to keep a reasonable supply of shares, together with their retaining bolts or pegs, such items as coulters, skimmers and mouldboards are usually only stocked on farms where the soil is very abrasive and/or where large acreages have to be ploughed. In these conditions, it may be wise also to stock disc-coulter bearings. Obviously, the maximum possible standardisation of plough types is an asset. Current trends suggest a greater degree of standardisation in future between different makes of plough.

Safety precautions

Although tractor ploughing is not particularly hazardous under normal conditions, a number of accidents do happen each year— some of them fatal. Here are some precautions:

1. Always exercise care when attaching the plough to the tractor. Some inexperienced operators have got caught up in the linkage system, as a result of operating the tractor hydraulic control lever while standing on the ground to attach the links to the cross-shaft. This practice is quite unnecessary.

2. Remember that when a mounted plough is in the raised position on the tractor linkage, the centre of gravity of the outfit is also raised considerably. The outfit is therefore less stable and liable to tip sideways on gradients or when turned abruptly at speed. A raised plough also makes the tractor steering 'light'.

3. Avoid the repetitive use of one independent brake during routine ploughing as this tends to unbalance the adjustment. In fact, the headland should be wide enough to eliminate the necessity for using the independent brakes at all for turning purposes.

4. Never kneel or sit under a plough held in the raised position by the tractor hydraulic system when fitting shares or making adjustments.

5. Never dismount from the tractor whilst the outfit is in motion. Several tractor drivers have been run over by their own machines in this way.

6. Exercise care when tightening or slackening nuts on coulter brackets which are positioned immediately above the sharp edges of the disc coulters. Such bolts must be securely tightened, and the use of strained spanners should be avoided because of the risk of their slipping.

7. Ensure that no one is standing near the swinging bodies of a stationary reversible plough before tripping the turnover mechanism.

Construction and operation of disc ploughs

Although disc ploughs are not very popular in this country—being regarded as unsuitable for the bulk of our soil types and tillage techniques—they could be used more than they are. They require less maintenance than mouldboard ploughs, particularly on abrasive and stony soils.

The construction of the disc plough, as shown in Fig 52, is relatively simple but very robust. The concave discs which cut,

59

elevate and turn the soil may range from 24 in to 36 in in diameter on various makes and models. The attitude of the discs can be regulated by altering their vertical and approach angles, and these factors regulate penetration, depth of working and movement of soil. The scrapers fitted to the concave faces of the discs, besides keeping the latter clean, can usually be adjusted to alter the throw

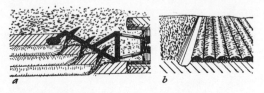

52 Disc plough: *a* At work, *b* Furrows produced

of the furrow. The pulverising action of the discs on the soil simplifies subsequent cultivating operations—an asset where these have to follow immediately. In stony, rooty land, the plough will ride over obstructions and avoid damage.

Against the disc plough, however, it must be said that weed control, and the burial of trash in some conditions, is not particularly good. Also, the firm packing of furrows, characteristic of the mouldboard plough, is lacking. Furthermore, when ploughing land which has been heavily traversed with tractors and harvesting machinery (particularly sugar-beet land), the disc plough's less positive action does not transpose the damaged surface soil as effectively as does the mouldboard plough.

As with many mounted mouldboard ploughs, the working depth of a disc plough is controlled by the tractor hydraulic system, and/or by a land wheel on the plough. The tendency of the plough to 'crab', due to the angle of the discs, is usually checked by the angling of a flanged rear wheel and sometimes by the angling of the land wheel. Excessive side-thrust can be checked by observing the slack on the tractor lower-link check chains— these should have an equal amount of slack on either side. The plough should be level both laterally and longitudinally when in work, and the necessary adjustments may be made with the usual tractor linkage levelling screw and top-link length respectively.

Ground speed influences both the quality of work produced and the levelness of the finish.

Penetration of the discs can be regulated on some makes of plough by utilising alternative cross-shaft and top-link positions. Some disc ploughs are designed to carry ballast weights in appropriate conditions. It is possible also on some ploughs to vary the furrow width by means of a lateral adjustment provided for each disc unit. Width of the front furrow is usually regulated by lateral positioning of the plough frame on the cross-shaft.

Construction and operation of chisel ploughs

The structure of a chisel plough resembles that of a cultivator more than that of a plough (Fig 53). It usually takes the form of a robust frame—rather like the standard tool bar designed for three-point linkage attachment—carrying from three to five rigid or spring steel tines capable of penetrating to a depth of 14 in or more.

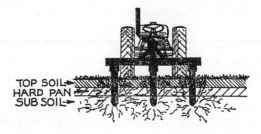

TOP SOIL
HARD PAN
SUB SOIL

53 Rear view of chisel plough at work and of its shattering effect on top soil and subsoil

Its primary purpose is to break up any hard pan which may have developed at the subsoil level as a result of the repeated traversing of machines during previous tillage and harvesting operations. It is a valuable implement for improving aeration and drainage, and soil conditions generally.

The optimum depth of working must depend upon the depth of the top soil and the nature of the subsoil. The ultimate operating depth is normally arrived at gradually by a number of treatments or *passes*. Operating adjustments are similar to those for the rigid tine cultivator described in Chapter 2. The provision of alternative tines and shares makes it a versatile implement. Its bursting effect is much greater if it is used in dry soil conditions, although of course its draught will then be at its highest.

Construction and operation of mole plough/subsoilers

The mole plough consists of a strongly constructed frame and vertical leg to withstand the inevitable strain imposed upon it (Fig 54). At the foot of the leg may be attached either a share or a cylindrical *mole* (Fig 55). This valuable dual-purpose implement is available in both mounted and trailed versions, and is capable of improving crop yields substantially by improving soil and subsoil conditions.

For the mole-draining operation, the subsoiling attachment is removed and the mole attachment fitted. The plough is drawn through the subsoil, and the bullet-shaped mole (Fig 55) creates a tube-like cavity about 3 in in diameter which, in a clay subsoil, is quite durable and acts most successfully as a drainage channel. The operation is usually performed in spring or autumn. The draught of this implement can be extremely high, calling for a powerful tractor. Some trailed types are designed for operating in conjunction with a tractor winch.

The direction, spacing and depth of the moling depends on the nature of the subsoil, the undulations of the land and the position and depth of existing tile drains and ditches. Where circumstances permit, a spacing of 9 ft between the moles is desirable, and in most cases a depth of about 20 in is most effective.

A vertical, steel disc in front of the plough leg minimises surface disturbance by making an incision in the soil or turf. In very hard conditions, however, it may be necessary to remove the disc to obtain adequate working depth.

The tube-like tunnel can be made larger and more effective by attaching an expander behind the mole, 4 in or more in diameter. A 4-in expander is sometimes used as a main drain to carry water from 3-in subsidiary moles. Such main moles should be 4 in to 6 in deeper than the smaller ones.

In carrying out the moling operation, the outfit should travel up-hill, i.e. against the fall of the land. Ideally, the movement of the plough should not be interrupted once the run is begun.

Needless to say, the mole-draining process is not satisfactory in gravelly or sandy subsoils as the cavity would very quickly collapse. In such cases, it is preferable to use the subsoiling attachment shown. Subsoiling is most effectively performed

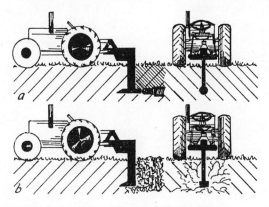

54 Mole plough/subsoiler: *a* Moling, *b* Subsoiling

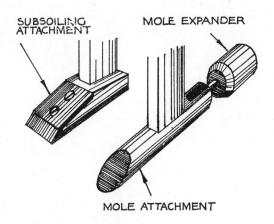

55 Mole plough/subsoiler fitments

immediately after harvest when drier conditions cause a greater bursting and shattering effect on the soil and subsoil. If the main aim is to improve the drainage, bouts should be made across the field at approximately 4 ft intervals. If, on the other hand, the object is primarily to break up hard pan, then the bouts should be 2 ft apart.

The mole may also be used to lay polythene tubing for water supply. Special moles are available with attachment points for the tubing.

Ploughs play a vital part in producing suitable soil conditions for seed germination and plant growth. On some light soils, the ploughing operation produces a lot of tilth, and only a little cultivation is necessary to prepare the seedbed. On some heavy soils, too, repeated ploughing is more effective than cultivating machinery in working down the soil. However, the wide range of soil types, the unreliability of the British climate, and the variety of crops grown in this country, are jointly responsible for the vast array of cultivation appliances in use today.

The aim of cultivation is to provide an adequate proportion of fine crumb-like soil particles, through which air and moisture can freely circulate. The texture and degree of compaction should be such that the seed is embedded in close contact with the soil, and at a consistent depth. Once the moisture and warmth of the soil has caused the seed to germinate, the soil should provide good anchorage for the plant roots.

The smaller the seed to be sown, the finer the tilth needs to be, but it is not always easy to obtain fine tilth, despite the wide variety of implements available. Indeed, it presents a serious and expensive problem for those who farm heavy clay land. Furthermore, the soil should be free from weeds, which would otherwise compete with the seedlings for plant nutrients, air and light. All these ideal soil conditions can be obtained only if the nature and humus content of the soil is right, and the various operations entailed can seldom be carried out without some undesirable effects such as the inevitable compaction of the lower layers of soil by the repeated passage of tractors and implements over the surface, and loss of soil moisture through the repeated exposure of fresh soil to the atmosphere. The latter is a matter of particular concern during late spring cultivations especially on the lighter soils.

When tilling soil, as in any other farming operation, one has to work in harmony with nature for satisfactory results. Mismanagement of land can add to the problems of *panning* (the compaction referred to above), *capping* (fine soil particles cementing together on the surface and impeding drainage and aeration), *smearing* (closure of soil air spaces by the friction of implements or wheel-spin in wet soil conditions), *baking* (the effect of hot sun and drying wind immediately following wet conditions on a clay soil, etc). Autumn and winter ploughing help to provide a good supply of tilth for spring seedbeds, but, even then, the timing of spring cultivating operations must be right. Soil types are so varied that each farmer has to select his implements, and time his operations, as judiciously as when harvesting. Use of the wrong implement, or even the right implement at the wrong time, can aggravate the problem and temporarily damage the soil structure. The wide variety of crops commonly grown—cereals, peas, beans, sugar beet and potatoes, to mention but a few—require differing techniques in seedbed preparation.

Spring and autumn seed bedsneed differing treatments. The latter, although demanding adequate tilth, is frequently left a little rougher on the surface than a spring seedbed, in order that the small clods may protect the emerging seedlings from wind and frost. The rougher surface also helps to reduce soil *capping*.

In addition to the immediate preparation of seedbeds, cultivation equipment is also used for stubble cleaning, the extraction of weeds and trash from the soil, the disturbance and destruction of soil pests, the integration of farmyard manure and artificial fertilisers into the soil, the levelling of the land surface in preparation for crops involving mechanised row-crop work and the spring scarifying of growing cereal crops and grassland.

It is important that implements are cleaned before they pass from one field to another, in particular where persistent weeds such as couch or wild oats exist, or even where soil-borne pests or fungi are prevalent.

Construction, operation and maintenance of tillage equipment

Rollers

Despite its simplicity, the agricultural roller more than justifies a place in any study of cultivating machinery.

The flat surface roller, once favoured on light soils, has been superseded on most farms by the ribbed or *Cambridge* type. This has a corrugated periphery, formed by the series of ribbed rings which make up the rolling cylinder. The ribs help to prevent capping of the soil, by leaving it finely corrugated. They also help to crush clods, where required.

The single roller, although commonly used, is not an economical load for the average tractor, and for this reason triple gangs of rollers have been developed (Fig 56).

The consolidating effect of a roller on the soil is governed by four factors: weight, diameter of rings, type and condition of soil, and speed of travel. The first two factors should be considered when selecting a roller, and the last two when using it. The roller has various applications, and some of the more important ones follow:

1. Rolling of newly ploughed grassland or clover ley. The criterion in this operation depends upon the time of year. In the autumn or winter, well-packed furrows are wanted to discourage growth of grass at the furrow joints. On spring or summer plough-ing, the furrows should be pressed down, in order to close the cavities which exist between the furrows and the subsoil, and prevent the crop, when sown, drying out through lack of subsoil contact. In the second case, rolling is more effective if, during the previous ploughing, the coulters have been tilted to undercut the furrow wall (Fig 28b). Special furrow-presses, designed for attach-

67

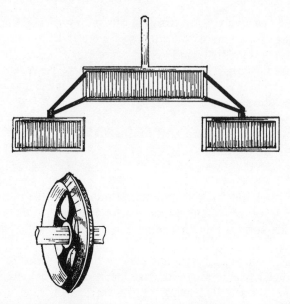

56 Triple-gang Cambridge rollers and typical ring

ment to the plough, have been used for this purpose in the past, but are now rarely seen. Any rolling intended to compact the soil should be done at low speed. Because it is a relatively light job, there is a tendency to go too fast. This temptation should be resisted, as even at low speeds the degree of compaction achieved is much less than one would imagine.

2. Rolling to break clods. This can be effective where soil conditions are such that a high proportion of clods have escaped the impact of cultivator and harrow tines. It should be done faster than when rolling for compaction; the aim is to break the clods by impact, rather than by vertical pressure, as the latter would tend to push them below the surface.

3. Rolling cultivated land. Sometimes, during cultivation operations, a seedbed becomes too *fluffy* or *hollow* and needs compacting. The advisability of rolling between cultivations or after the seed-drilling operation must be carefully considered. Sugar beet seedbeds are sometimes rolled before drilling, to provide the

smooth surface and firm conditions necessary for unit type seed drills.

4. Rolling of growing crops. This is often desirable in the spring, to compact the soil around the plant roots, particularly those of autumn-sown cereals, after winter frosts have loosened the soil and lifted the plants. Rolling is sometimes used as a means of restricting wireworm movement and for conserving moisture.

5. Rolling-in grass seed. The use of a ribbed roller is a convenient method of introducing fine grass and clover seed into the soil. In some cases, the land is simply rolled in one direction, sown simultaneously by a seeding attachment fitted to the roller, then subsequently cross-rolled at right angles to the first direction.

6. Rolling of grassland and clover ley. This is another spring operation, usually carried out on crops scheduled for mowing, especially temporary leys, where the soil may benefit from rolling after winter frost. Where the land is stony, rolling reduces the risk of damage to mowing machine cutter bars and forage harvester flails.

OPERATION

Reference has been made to the effects of ground speed, where this is of particular importance. Other considerations are the avoidance of fast and abrupt turning on growing crops, and the importance of stopping work if rain moistens the soil sufficiently to cause it to stick. Newly ploughed furrows should be rolled in the same direction as the ploughing. In other circumstances, it is often better to roll round and round the field following the boundaries. When transporting the roller to and from the field, avoid excessive speeds or rough and stony tracks, as sudden impact on the rings might crack them.

MAINTENANCE

Various types of bearings are used on roller spindles with differing lubrication requirements. Some are totally enclosed ball or roller bearings, requiring only occasional attention, while others are of brass or bronze, and demand fairly frequent greasing. The spindle should not be lubricated at the hubs of the actual rings, as the lubricant gathers dust, which causes excessive wear.

After a period of service, the rings develop end-float. Sometimes, the only way of taking this up is to withdraw the central spindle from one end sufficiently to insert packing washers between the side frame of the roller and the end ring. On other rollers, either the bearings or the side cheeks can be slid towards the rings.

Tool bars

Several cultivating operations which once required separate implements can now be carried out with a range of attachments designed to fit on to a universal tool bar attached to the tractor three-point linkage. In the following pages, for simplicity, the tool bar is considered an integral part of such implements. The tool bar itself is described under *Inter-row cultivating equipment*.

Rigid-tine cultivators

The rigid-tine cultivator is probably the most common of all cultivating implements. As its name suggests, it comprises a number of rigid tines, mounted on a robust frame. The trailed version has been a popular tractor-drawn implement, which still has the advantage over most of the newer mounted types, of a greater longitudinal clearance between the rows of tines. Fig 57 illustrates the trailed type. It will be seen that the tines are staggered to achieve intensive cultivation while avoiding choking by clods of soil, stones or trash. The tines can sometimes be arranged in pairs or groups for row-crop cultivation (Fig 58). Alternative types of shares (or shovels, as they are sometimes called) are available, some of which are reversible, having two cutting edges to provide a double lease of life (Fig 59). The advantages of changing shares to suit the conditions, however, are seldom made full use of. In heavy working land, or when penetration is difficult, the narrow shares should be used; but in easier working conditions it is better to have the extra cutting width of broader shares.

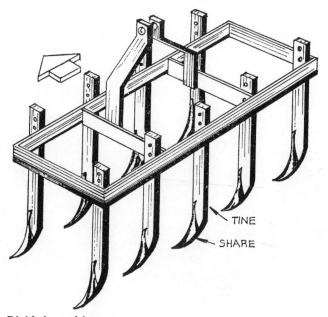

TINE

SHARE

57 Rigid-tine cultivator

58 Inter-row cultivating

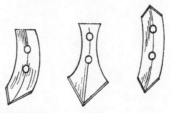

59 Types of cultivator share

OPERATION

When a trailed cultivator is attached to the tractor drawbar, the hake adjustment, i.e. the vertical positioning of the hitch point, should be set so that the frame of the cultivator is level with the ground. Working depth is varied by a screw lever, which operates the cranked land-wheel axle. The depth can sometimes be varied further by raising or lowering the tines in their supporting brackets. In hard working conditions, it is advisable to raise the tines for greater rigidity.

With mounted-type cultivators, the pitch of the tool bar is regulated by varying the length of the top link. The depth is controlled either by adjustable land wheels, or by the tractor hydraulic control.

The rigid-tine cultivator is primarily a deep-cultivating implement, although it is sometimes used for the shallow scarifying of stubble after harvesting operations, to create sufficient tilth for weed seeds on the surface to germinate, so that they may be destroyed by subsequent ploughing. On some cultivators, the tines, although rigid in themselves, are pivoted and spring-loaded to avoid damage if they strike an obstruction in the soil (Fig 60). A slight reduction in draught has been claimed for this type of cultivator, due to its resilience in operation.

MAINTENANCE

This implement demands very little attention, but on trailed cultivators, the land wheels and trip mechanism should be kept lubricated. Share and tine wear may be considerable in abrasive soils. When necessary, new shares should be fitted, or double-ended type reversed, otherwise damage to the tine itself will result. Undue side play, at the pivots of spring-loaded tines, should be taken up with packing washers.

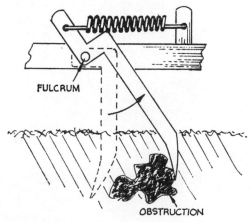

60 Spring loaded tine

Spring-tine cultivators

Confusion occasionally arises between spring-tine cultivators and the spring-loaded tine type referred to above. Whereas the spring-loaded tine is designed primarily as a mechanical safety factor, the spring tine is designed for its effect on the soil. The spring steel tines help to pulverise the soil by their vibratory action (Fig 61).

61 Spring steel tine

The main structure of the cultivator may either employ the usual mounted tool bar, or be constructed as a separate implement, designed for three-point attachment (Plate 5).

Working depth is controlled either by depth wheels or by the tractor hydraulic system. On those types having no land wheels, a stabiliser fin restricts lateral movement of the cultivator, when in work (Fig 62). The curvature of the spring tines raises couch or

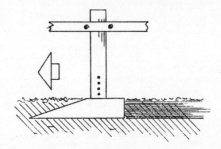

62 Stabiliser fin

other trash to the surface, for collection in soil-cleaning operations. Sometimes, however, this elevating effect may be undesirable, as for instance when long farmyard or green manure has been intentionally ploughed in, or when in late spring cultivations moisture conservation is important. In some conditions, too, this cultivator tends to bring clods to the surface, which may be contrary to what is required. It is therefore important to regulate working depth and the disposition of the tines carefully.

Maintenance of the spring-tine cultivator is similar to that of the rigid-tine type.

Rotary cultivators

Fig 63 illustrates a typical rotary cultivator. Both tractor-mounted and trailed types are available, the former being more manœuvrable at corners and in confined spaces, while the latter is simpler to attach, and imposes fewer stresses on the tractor when in work. In either case, the power is supplied by the tractor power take off (*p.t.o.*) to the rotor shaft, which carries a series of blades or knives designed to chop and throw the soil. A slip clutch, or cushioning device, is incorporated in the rotor drive, to prevent serious damage if the blades should strike an obstruction. The power-driven cultivator can operate in difficult soil conditions, because the rotor revolves in the same direction as the wheels of the tractor, thus assisting traction of the outfit. However, whether

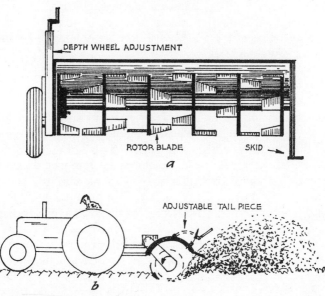

DEPTH WHEEL ADJUSTMENT

ROTOR BLADE   SKID

a

ADJUSTABLE TAIL PIECE

b

63 Rotary cultivator (p.t.o. driven): *a* Rear view of rotor and blades, *b* Rotor action

or not the machine should be used in any particular condition requires careful consideration, and is sometimes a subject of controversy. The rotary cultivator is excellent for forcing a tilth in stubborn soil, but indiscriminate use can result in hollow, fluffy soil conditions, which are not always desirable. In some unfortunate cases, where heavy land has been left in this condition, subsequent rain has turned the fields into a quagmire. It should also be realised that soil structure can be damaged by over-cultivation with a machine of this type.

75

It will be seen that, wisely employed, the rotary cultivator is a boon, but untimely use can be detrimental. The machine is favoured by some farmers for the preparation of potato land. It is also used for stubble cleaning, by damaging growing weeds and producing tilth to induce further weed seeds to germinate, so that they may be destroyed by a later treatment or by ploughing. It aerates the soil well, and controls rhizomatous weeds effectively, so long as weather conditions and management allow the repetition of the treatment at the right times.

Small versions of the rotary cultivator are obtainable for use in gangs for row-crop work.

OPERATION

As the rotor is driven by the tractor p.t.o., its speed is related to that of the tractor engine. Thus, changing the tractor gear regulates the intensity of the cultivation by altering the ground speed of the outfit. Also, some machines are equipped with a gear box enabling alternative rotor speeds to be selected, and this increases their operational flexibility. The production of the fine powdery tilth, desirable for some spring-sown crops, generally requires a high rotor speed relative to the ground speed of the outfit, and the tail flap should be lowered, to achieve a shattering effect as the soil is thrown against it. The rougher seedbed requirements of autumn-sown cereals, on the other hand, necessitate a lower rotor speed and the raising or removal of the tail flap.

Rotovating is an effective method of introducing farmyard manure, crop residues, fertilisers and insecticides into the soil, and settings for these operations are best found by experiment.

Working depth is usually regulated by a land wheel and/or skid. In heavy land, or unploughed soil, it is usual to work down to the ultimate cultivating depth in stages, the first treatment reaching a depth of about 4 in, and subsequent ones going deeper, until the desired depth is reached. The alternative systems of working in the field are either in lands (similar in principle to the method used for ploughing with the right-hand plough) or round and round.

MAINTENANCE

The rotor bearings normally require periodic greasing, but, before applying a grease gun, be sure to remove the soil which invariably covers the nipples. The grease nipples on the power-drive shaft universals, and the land-wheel bearing where fitted, also require regular attention.

The blades of the rotor should be checked frequently for condition and tightness—their rate of wear is naturally higher in abrasive soils. They are usually held by two bolts, and one of these sometimes has shear-strength lower than the other, so that, if an obstruction is hit, the effect of the impact is reduced by one bolt shearing and the blade hinging on the other. The machine should not be operated with blades missing from the rotor, as this upsets the balance of the latter, besides causing intermittent loading on the rotor-drive mechanism. After a period of storage, it is advisable to check that the slip-clutch or cushioning device has not seized. The bevel gear box and the rotor-speed gear box generally have oil baths which need periodic checks on the oil level; a change of oil is sometimes recommended at the end of a season. The chain drive, between the counter-shaft and the rotor shaft, also requires lubrication and periodic checks on the chain tension.

Disc harrows

Basically, the disc harrow comprises either two gangs of con-cave discs in tandem (Fig 64), or four gangs in tandem abreast

64 Two-gang tandem disc harrows

(Fig 65). The discs may range from 16 in to 24 in in diameter, and the gangs are free to rotate by ground contact. The discs cut the soil and throw it aside. By angling the gangs diagonally, and giving the concave faces of the discs a slight forward inclination, penetration is achieved. Increasing the angle increases the depth of penetration. The gangs have a limited amount of freedom for vertical movement, so that they can operate effectively on undulating land. Both mounted and trailed types are available, the former being more manœuvrable, but limited in size (and so in

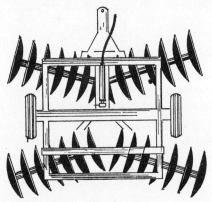

65 Four-gang tandem abreast disc harrows

working capacity) by the restriction of weight overhang on the tractor when in the raised position.

Transport of the trailed type necessitates the use of wheel assemblies which can be quickly attached. Unfortunately, some of these have been rather shabbily designed, and are somewhat unreliable. A recent improvement is the provision of solid rubber or pneumatic-tyred wheels. On some models, the same wheels also have an operational function, giving control over working depth, and ease of turning on headlands (Fig 65).

OPERATION

The weight of the implement is matched with the size of its discs to achieve good penetration, although brackets are sometimes fitted to carry additional weights, if desired. This obviates the common makeshift practice of loading large and ungainly objects on the implement, a practice which may hamper the rise and fall of the gangs on undulating land.

Besides producing tilth effectively in difficult conditions, the disc harrow also achieves a certain amount of compaction of the seedbed, which may, or may not, be desired. It excels in producing seedbeds on newly ploughed grassland, where the soil needs to be worked down without disturbing the furrows or bringing the grass to the surface. In this case its consolidating effect is also valuable. Its surface action is an advantage in the spring, when the weathered tilth should be retained on top of the seedbed. A less frequent, though nevertheless valuable, application of the disc harrow, is the cutting down of surface flag or green manure, before ploughing in.

It should not be used on land infested with couch, docks, or similar persistent weeds, as the cutting action of the discs would increase the weed population.

To ensure a level surface, the gangs must be correctly set. The angling of the disc gangs may be controlled hydraulically, manually, or by the draught of the tractor to a pre-set position. On some double-tandem disc harrows the two rear gangs have a limited adjustment, independent of the front gangs. On mounted-type single-tandem disc harrows, the top link must be set at the correct length, or the front and rear gangs will operate at unequal depths; with single tandem types this would result in side-draught of the implement, and a constant pull on the steering of the tractor. Scrapers are usually fitted close to the concave face of each disc to prevent soil accumulating and blocking the implement. If the scrapers are disturbed, they should be re-set as close as possible to their respective discs, without rubbing unduly. On some makes a pivot enables them to be swivelled across the face of the disc by the tractor driver, from his driving position, with the aid of a connecting cord. The quality and levelness of the work is influenced by the speed of travel.

D

MAINTENANCE

Disc harrows fitted with wooden, brass, or bronze bearings need frequent greasing, while those fitted with sealed bearings require little attention for long periods of service. Indeed, care must be taken with such bearings not to over-grease, or to force grease into the bearings with excessive pressure, as this can damage the retaining seals. Lubrication of the links connecting the front gangs to the rear should be avoided, as oil and grease attract abrasive soil, and so accelerate the rate of wear.

With a new implement, it is important after the first few hours of work to check the tension on the nuts securing the discs and distance pieces to their central spindles. Whenever these are tightened, remember to relock the nut by means of the cotter-pin or tab-washer provided. If the discs develop excessive end-play, their central holes will be enlarged, and the spindles damaged, making tightening of the disc assemblies impossible without re-conditioning. When the implement is new, the rear gangs are arranged so that their discs cut the soil in between the cuts made by the discs of the front gangs. This staggering effect intensifies the operation and the setting should therefore be maintained. It is likely to be upset by the distortion of the connecting links or supporting stays, resulting from careless reversing when the implement is in the soil, or from shunting the outfit when turning on the field headlands.

Tine harrows

Tine harrows are available in a wide range of sizes and weights and with various patterns of tines. The spring-tine type (Fig 66) is like a small version of the spring-tine cultivator, and its principle of operation is similar. The main difference is that the depth of working of the cultivator is controlled by raising or lowering the implement vertically, while that of the spring-tine harrow is usually controlled by varying the attitude of the tines. This is achieved by means of a lever connected to the cross bars which carry the tines. By rotating the cross bars, it is possible to regulate

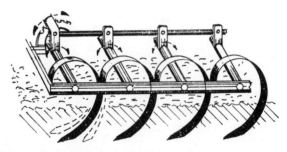

66 Section of a spring tine harrow

the penetration of the tines and their lifting effect on the soil. This implement is one that has stood the test of time, and is still well favoured in this day of advanced mechanisation.

Another widely used type of harrow is the straight-tine version, made in a zigzag pattern. As shown in Fig 67 the tines are arranged in several rows, and staggered to give an intensive harrowing effect,

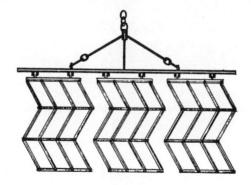

67 Straight tine zig-zag harrows

without causing too much restriction to soil clods and trash. The length of tine, which may range from 3 in to 8 in, is related to the overall weight of the harrow. Medium-weight harrows are widely used in the final stages of seedbed preparation, as these not only break down some of the smaller clods, but also help to level the surface.

Lighter harrows, of this type, are used to harrow in fertilisers before drilling, and seed after drilling. *Drags*, as the larger and heavier versions are sometimes called, are used to prepare seed beds for crops requiring a greater depth of fine soil. Other uses include collecting crop residues from the surface of the soil, and dragging out potatoes left behind the digger.

Some harrows are fitted with tines which are flattened and slightly curved in a forward direction (Fig 68). As mentioned in

68 Two common types of harrow tine

connection with the curved-spring tine, the effect of this may not always be desirable, especially where trash or clods are beneath the surface, or where it is important to conserve moisture.

Most harrows are comparatively light in draught, and often sets of eight can be drawn behind a tractor without overloading it. A set of three or four lightweight, or even mediumweight, harrows is not an economic load for the average tractor. In large fields, a set of harrows may be trailed behind a cultivator, or a roller, to load a tractor more efficiently. Combining two operations in this way also helps to reduce the soil-moisture evaporation which occurs between separate operations.

With the zigzag pattern harrow, now in common use, each unit slightly overlaps its neighbour, and has complete flexibility to rise and fall over ground undulations.

Some harrows are harnessed to a frame which is attached to the tractor three-point linkage. This is an advantage for transporting, turning at ends of bouts, and, often more important, for discharging rubbish from the harrows whilst in motion, by lifting them out of work. Mounted harrows retain their flexibility by the use of chains or flexible couplings between the harrows and the lifting frame.

Harrowing can be carried out round and round, or in bouts parallel to one field boundary. If the latter method is used immediately before drilling, the bouts should be at right angles to the proposed drilling direction, so that the drill marker-furrow can be easily seen.

Special small harrows are available for row-crop work, and some are arched for harrowing potato ridges—these are sometimes referred to as *saddle-back* harrows.

The only maintenance required by harrows is that of keeping the tines secure in the frames, and replacing them when they become excessively worn.

Chain harrows

Chain harrows are constructed in the form of a chain link mat, which is completely flexible to follow ground undulations (Fig 69a). Some versions have plain links, one example of which is shown in Fig 69b, and these are commonly used on grassland—particularly

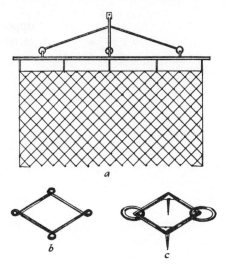

69 Chain harrow

that scheduled for mowing. The harrow spreads manure left on the surface by stock from the last grazing period, and invigorates the sward by its brushing action. It is sometimes used on arable land to roll couch or other trash, which has been worked up to the surface of the soil for collection.

Fig 69c shows another type of link having spikes on its underside. The spiked links are more severe in their action. The incisions aerate the sward and help the introduction of fertiliser, particularly if the ground is harrowed immediately before the fertiliser is applied.

Some spiked and tined chain harrows can be inverted so that the smoother side may be used for less severe treatment. A further innovation is the fitting of scrapers on the front links, to disintegrate dung and molehills more effectively.

Weeders

Weeders may not be one of the most popular implements, but on some soils, and in certain crops, they are most valuable. As shown in Fig 70, this implement has many closely spaced spring tines,

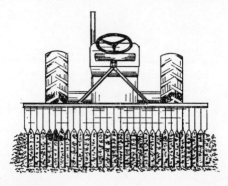

70 Weeder (rear view)

which lightly scarify the surface of the soil supporting growing crops. The shallow vibrating action of the tines rakes out weeds in their early stage of growth, and, provided the operation is correctly timed, no serious damage is done to the crop. The secret of success lies mainly in the knowledge of when to use the weeder, and this is usually when the crop is established, but while the weeds are still susceptible to the scarifying action. Another use is for scarifying a 'stale', prepared seedbed, immediately before drilling.

A weeder can often be used to advantage on winter wheat in the spring, on potato ridges until the crop has up to 4 in of foliage, and on peas as soon as the crop is established. For the last application, the treatment should be in line with the pea drills. The operator must use his discretion in these operations, particularly the last two, to determine how severe a treatment is needed. This is controlled by the angle of the tines, which can be adjusted by altering the length of the top link (assuming the implement to be of the more common, mounted type). At the two extremes of this adjustment, the tines can be made to brush the soil lightly or scarify it severely (Fig 71).

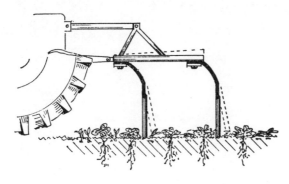

71 Weeder (side view)

Small versions of the implement are available as attachments for corn drills. The action of the tines covers the seeds, and eliminates the need for harrowing.

Unclassified implements

The many complexities brought about by varying soil types and conditions, the fickleness of our climate, and individual crop requirements, have led to the development of many machines designed to simplify soil cultivation. Some of these implements are never produced in quantity, others enjoy only a short period of popularity, and still others appear to be favoured only locally. The better known and more promising types include ground-driven rotary tillers (Fig 72), reciprocating harrows (Fig 73), variable-pitch and self-cleaning harrows (Fig 74), helical cultivators (Fig 75), scrubbers and levellers (Fig 76).

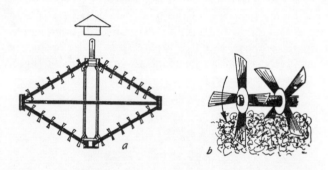

72 Rotary tiller (ground driven). *a* Layout of tiller, *b* action of tiller blades

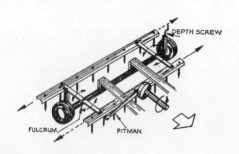

73 Reciprocating harrow

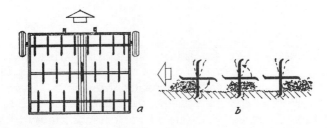

74 Variable pitch self-cleaning harrow

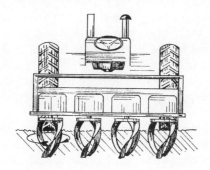

75 Helical cultivator

76 Land leveller and clod crusher

Inter-row cultivating equipment

Because of their size and nature, certain farm crops, such as potatoes, sugar beet, mangolds and turnips, require a greater area for their growth and development than others. The roots need more soil, and the foliage needs more space and light. Unfortunately, when plants are spaced out, the cultivated soil between them invites the germination of weed seeds, particularly in the early stages of the crop's growth. These weeds, if not subdued, compete with the crop for light, air, moisture and plant foods, thereby seriously depressing the ultimate crop yield. In order to simplify inter-plant weed control and periodic cultivations, seed is sown, or the plants are set, in evenly spaced rows.

Inter-row cultivating is another sphere of crop production which has been simplified by the advent of the tractor hydraulic system of implement attachment and control. Row spacings used for some crops have been slightly modified in recent years to suit the use of general-purpose tractors, as opposed to specially designed row-crop tractors with narrow wheels.

The soil-working attachments for row-crop cultivation are often fitted to the standard rear-mounted tool bar described earlier in this chapter. Other arrangements use a mid-mounted under-slung tool bar or a front-mounted tool bar.

REAR-MOUNTED TOOL BAR

For work which does not call for extreme precision i.e., where it is not necessary to hoe close to the plants, the normal tool bar, with check-chains tightened to stop side-sway, can be used (Fig 58). When the tool bar is fitted with a stabiliser fin, it is possible to achieve considerable accuracy of inter-row work (Fig 62).

Close hoeing, however, which is required with many root crops, particularly in their early stages of growth, involves a second operator, and a steering mechanism embodied in the tool bar and its linkage (Fig 77). The tractor driver concentrates on steering the outfit down the rows, while the tool bar operator rides in the seat provided, and steers the tool bar accurately in relation to the drills, compensating for any minor deviations in the latter. Although two men are involved, it is often claimed that the extra labour cost is more than offset by the increased output, resulting from the sharing of the concentration.

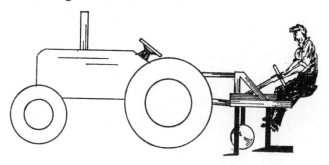

77 Rear-mounted, steerage type tool bar

MID-MOUNTED TOOL BAR

For one-man row-crop work the mid-mounted (or underslung) position of the tool bar (Fig 78) is probably the most satisfactory.

78 Mid-mounted tool bar

It is clearly visible to the tractor driver, and its location in the centre of the tractor reduces the sensitivity of steering reaction. This is in contrast to the exaggeration of steering movements suffered by front and rear-mounted tool bars.

The main drawbacks of the underslung tool bar are the height restrictions, imposed by the rather limited ground clearance of the transmission housing of most tractors, and the inaccessibility of the central hoes on the tool bar.

FRONT-MOUNTED TOOL BAR

With the front-mounted bar, the tractor driver does not have to be continually looking backwards to watch the work being done, but there are always some of the hoe units out of sight. The steering is very critical, and in some cases heavier than normal, due to the weight of the tool bar on the front axle of the tractor, although, on some tool bars, castor wheels are provided to carry this extra weight (Fig 79).

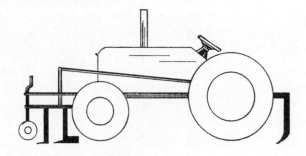

79 Front-mounted tool bar

ATTACHMENT AND PREPARATION OF TOOL BAR FOR WORK

The tractor wheel-track width (measured between tread centres) should be a multiple of the crop-row width. For many root crops, this is three times the row width; for example, a crop of sugar beet with 20 in row spacing, would need the wheels of the tractor to be 60 in apart (see Fig 80). The method of setting the wheels varies from tractor to tractor, so that reference should be made to the operator's instruction book, which indicates the necessary wheel-centre and rim dispositions to give the required setting. Although the front tyres of the tractor are considerably narrower than the rear ones, it is usually necessary to alter their position as well, in order to avoid damage to the plants.

For one-man hoeing with a mid-mounted tool bar, driving may be simplified by fitting a pointer on the front axle, to line up with a row of plants in the driver's line of vision.

Any necessary adjustment of check-chains, or fitment of stabilising stays should be carried out when attaching the rear-

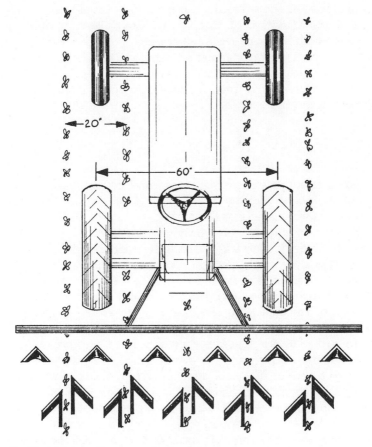

80 Positioning of tractor wheels and hoes

mounted tool bar. Generally speaking, with one-man hoeing, the check-chains or stabiliser are set to prevent side-swing of the tool bar whilst in operation. With *steerage* hoes, however, some flexibility is desirable to permit the independent steerage of the tool bar.

If alternative top link connecting points are provided on a rear-mounted tool bar, the uppermost one gives greatest sensitivity for the tractor's hydraulic draught control. Sometimes, a spring is fitted at the top link connection points, to increase the sensitivity of the draught control through the top link. In any event, top-link length is critical to obtain a consistent depth of working between both front and rear rows of hoes.

The choice, positioning, and setting of the attachments on the tool bar depends on the nature of the crop, its stage of development, and the general working conditions. Let us consider these factors in detail.

Hoeing attachments

Hoeing attachments are tines to which hoe-blades are attached. Many different patterns of blades are marketed, but those in most common use are broadly classified as *A* hoes and *L* hoes by reason of their shape (Figs 81 and 82).

A further type of attachment sometimes used is the disc assembly. Fig 83 shows a pair of *L* hoes and a pair of discs in relation to a row of plants. This is known as a gang arrangement. Each gang can rise and fall individually to accommodate undulations and maintain accurate depth of hoeing. The gangs are controlled by an adjustable tension-spring, to ensure penetration without bouncing.

In order to set these hoeing attachments efficiently, the operator must understand the precise function of each of them. The main aims of after-cultivation are weed destruction and aeration; these must be achieved without damaging the plants or smothering them with soil. The following points should be borne in mind:

1. *A* hoes should be spaced to travel midway between the rows of plants (Fig 80), and set deep enough to cut any weeds present

and to loosen the upper layers of soil. It is preferable, in level soil, to place the *A* hoes in front of the *L* hoes as this reduces the strain on the *L* hoes which have to perform the more delicate close hoeing. However, this results in tilth being moved away from the plants, which is not desirable where the soil tends to cap.

2. *L* hoes require more precise setting, as their function is to hoe the soil immediately adjacent to the plants, without damaging foliage or roots. When hoeing, the tractor driver or tool bar operator normally steers the tool bar by concentrating his attention on one pair of *L* hoes. It is advisable to have this pair set with slightly less margin on each side of the row of plants than the others. This ensures a safety margin on those not in the line of vision.

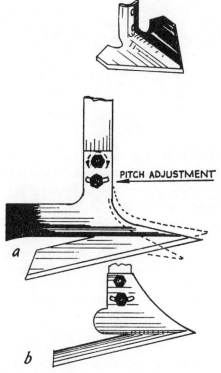

81 and 82 Types of hoe blade

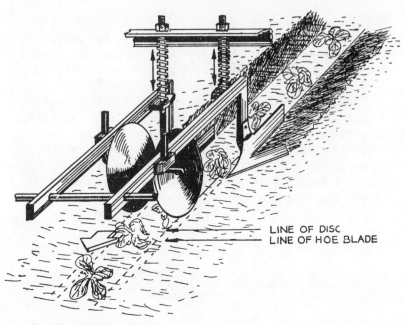

LINE OF DISC
LINE OF HOE BLADE

83 Hoe and disc setting

Depth of working is extremely important, to ensure that the weeds are severed from their roots; if too deep, the weeds may be left intact and merely raised to the better soil conditions created by the operation. Also, excessive depth may partially bury the crop plants, and this can seriously retard their development in their early stages of growth. There is also the danger of bringing up weed seeds from beneath the surface where they lie dormant to a level more conducive to their germination and growth.

When capped soil conditions are encountered, it may be difficult to achieve and maintain penetration of the hoe-blades. It may help to sharpen the hoe-blades, and, where possible, to increase their *pitch*. Some tool bars, however, are equipped with a hydraulic ram to provide downward pressure when necessary.

For accuracy in inter-row hoeing, it is vital that the number of sets of hoes on the tool bar matches the number of coulters or seeder units used for the drilling operation. This ensures that the tool bar never straddles a joint in the drilling bouts, as slight drilling discrepancies are bound to exist, no matter how skilfully the work is carried out.

Where a stabiliser fin has been used during the drilling of the crop, this same device may be fitted to the tool bar in such a position that it travels in the same groove, so ensuring accuracy of hoeing.

3. Discs are an advantage on certain soils, but they are only used in conjunction with A and L hoes in the early stages of the crop's growth. Their slight angle to the direction of travel provides a delicate hoeing action, very close to the plants, without damaging them. Discs eliminate the need for close hoe setting and its attendant problems of either cutting or smothering the seedlings. Because of this, slightly higher ground speeds are permissible. If set too deeply, the discs tend to move the soil away from the plants, leaving them rather insecure on a narrow plinth of soil, and vulnerable to the extremes of heavy rain or drought (Fig 84).

84 Effect of excessive disc depth

The discs should be set just deep enough to maintain a consistent, but shallow, hoeing effect on the soil surface. Their efficiency is seriously affected by any unevenness of the land surface.

DOWN-THE-ROW THINNERS

Given reasonable soil and weather conditions, the problems of mechanical inter-row hoeing are fairly easily surmounted. Another problem, however, which follows in the wake of hoeing, is that of thinning out the plants in order to achieve the optimum plant population, with, at the same time, satisfactory spacing of the plants within their rows. On a small scale, this work can be done by hand without undue hardship. However, on many large farms, with a big acreage of root crops, particularly sugar beet, these operations are a serious problem. One factor is their urgency, for a short delay in the thinning of a crop can seriously depress the development and ultimate yield of the crop.

Several systems have been devised, as the result of experiments, to reduce this problem, employing various types of equipment. One method is down-the-row thinning. Fig 85 illustrates a

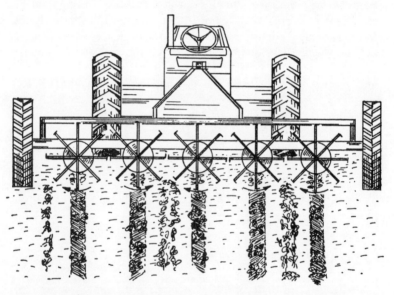

85 Down-the-row thinner

typical machine. Its working principles are easy to follow; its land wheels drive several rotating heads, each of which can carry a variable number of hoe-blades. The machine travels down the rows of plants with its hoeing heads set to match the row spacing. As the heads rotate, the hoe-blades make strokes at regular intervals across the rows, eliminating any plants in their paths

(Fig 86). The heads are contra-rotating, i.e. some of them rotate in a clockwise direction and others in an anti-clockwise direction; this prevents lateral swing of the implement when in work.

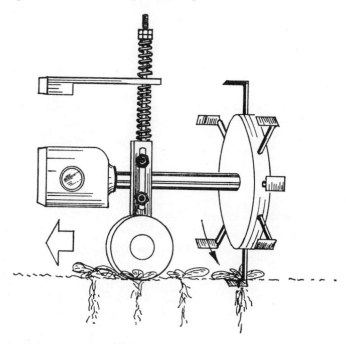

86 Action of spinner head

The lateral setting of the spinner heads is critical. It is usually best to offset the head so that the weeds are severed from their roots and lifted by the *upward* swing of the hoe-blades (Fig 87).

From one to three *passes* may be made through a crop, with *counts* being taken to assess the plant population before each treatment. The intensity of the operation can be regulated, either by using a different number of hoes on the rotating heads, or by varying their speed in relation to the ground speed of the outfit.

It will be seen that the system is operated on a purely mathematical basis; the crop loses the benefits of the selection of dominant plants, and the adjustment of plant spacing, both of

97

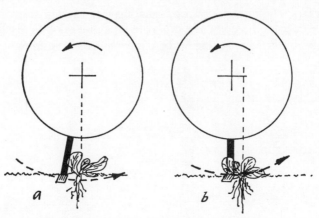

87 Alignment of spinner heads: *a* Central, *b* Offset

which are normally achieved by manual thinning. However, these disadvantages may well be offset by the saving in labour costs, and by the ability to thin out at the best time. The success of this system, in the beet crop, is usually enhanced by the use of so-called 'rubbed' and graded seed, sown with precision-spacing drills (see *Unit-type seed drills* in Chapter 4). Each rotating head is free within limits, under the influence of a control spring, to rise and fall and adjust itself to slight surface undulations, but in spite of this, the efficacy of the operation is greatly influenced by the levelness of the land. The penetration of the hoe-blades can be regulated, and is normally restricted to a depth of $\frac{1}{4}$ in to $\frac{1}{2}$ in. The control spring also helps to keep the hoeing depth constant.

There are many different types of row thinners, and Fig 88 shows one in which the hoe works with a pendulum action.

A recent development is the combination of both hoeing and row-thinning attachments on one tractor, to reduce wheel marks on the soil and to save time. However, this procedure demands accurate timing of the double operation.

GAPPERS

Gapping is sometimes carried out as an alternative means of

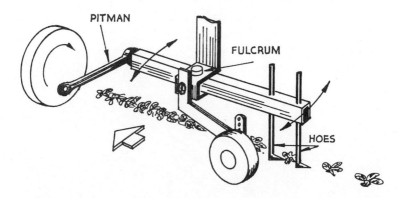

88 Principle of reciprocating thinner

reducing plant population. It may be performed with a down-the-row thinner having groups of hoe-blades removed from the rotating heads, or with specially designed rotating gapper-blades. It reduces the handwork necessary, but enables a certain amount of seedling selection to be made during the subsequent hand singling.

RIDGERS

Ridgers are available in both trailed and mounted versions, although comparatively few of the former are now used. The mounted type usually consists of the universal mounted tool bar, carrying three ridging bodies in line (Fig 89) or staggered, with the centre body positioned either slightly forward of the outer bodies or slightly behind them. The ridger is used mainly to form ridges for the planting of potatoes, to split the ridges after planting, and to mould up the growing crop. The usual bolt-hole drillings on the tool-bar frame enable the body positions to be adjusted to suit the particular row spacings for different varieties of seed and times of planting.

Each body consists of a vertical supporting stalk, carrying a narrow share and two mouldboards. The mouldboards of the central body are always adjustable, and on some makes the outer

99

ones can also be adjusted. When attaching to the tractor the top link must be carefully set, especially when the centre body is positioned either forward or rearward of the outer bodies. Insufficient pitch, i.e. excessive length of the top link, will accelerate the wear of the mouldboard ends, and result in rounded furrow trenches. A vee-shaped effect is preferable, to locate the tubers in

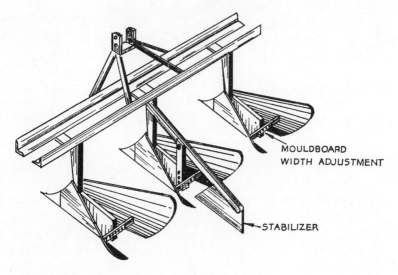

MOULDBOARD
WIDTH ADJUSTMENT

STABILIZER

89 Tool bar fitted with three ridging bodies

a regular line. Depth is regulated by land wheels, or by the tractor hydraulic control, as with other cultivators.

For good ridge formation, the ridger depends on thorough preliminary cultivation, at least to the intended planting depth. The last cultivating operation prior to ridging should be carried out across the intended direction of planting. This distributes the soil more evenly, and lessens any tendency of the ridger to wander; the line made by the marker is also more distinct. On totally suspended ridgers, i.e. those having no wheels, the steerage fin shown is necessary to stabilise the ridger, and obviate side-swing, especially on hillside land. The depth of the fin must be set by trial and observation; if too deep, it will prevent the ridger penetrating to the required depth; if too shallow, it will be ineffective. The use of land wheels simplifies the control of side-swing.

One maker provides an attachment consisting of two concave discs on a stalk; these are used with the concave faces together when planting, and the convex faces together when moulding up, so that the rims straddle the ridges.

Markers should always be used. Where the tractor track width is less than two crop-row widths, the distance between the outside body and the marker should be equal to one row width, *plus* the distance between the same outside body and the line taken by the appropriate front wheel of the tractor.

Where the track width exceeds two crop-row widths, the distance between the outer body and the marker should be equivalent to one row width, *minus* the distance between the outer body and the line taken by the front wheel of the tractor. The marker should be set deep enough to produce a clearly defined guiding line. The driver will need to keep the appropriate front tractor wheel immediately on the mark produced.

Work in the field usually commences with the drawing of a headland mark parallel with the boundary, made by tilting the ridger so that one body only, cuts a shallow and narrow marking groove in the soil. Either of the outer bodies may be used, although the one chosen will dictate the direction in which the outfit travels around the field, as the tractor wheel marks should be kept outside, i.e. off the land to be ridged.

When the direction of the actual ridging has been decided upon, a marking pole (or poles) should be set up, and the first run made with the ridger levelled up, but set very shallow, so that any steering discrepancies or ridger side-sway may be corrected on a return run down the same course. Once this opening bout has been completed, further ridging is comparatively straightforward, using a three-body ridger to produce two ridges at a time. Although this means one body is running light on each bout, the wheels on one side of the tractor will always be running in a trench, so keeping the row width constant.

Splitting the ridges, after the tubers have been planted, needs care and accuracy to ensure good coverage. Some operators prefer to split only two ridges at a time with a three-body ridger. After the first run, this avoids having to position the tractor on top of the ridges. Alternatively, the difficulty of three-row splitting can be reduced, by the use of spool-shaped guide rollers, fitted to a front-mounted tool bar.

MAINTENANCE OF ROW-CROP EQUIPMENT

The principal enemy of all forms of cultivating machinery is abrasive soil. This can cause severe wear on the soil-engaging parts and also on spindles and bearings. Most of the advice given in respect of the implements mentioned earlier is equally applicable to inter-row cultivation machinery.

Safety precautions

1. *Power-driven implements*
   Ensure all guards are in place.
   Do not make adjustments to moving parts or clear blockages while the machine is running.
   Always disengage the p.t.o. drive when work ceases, even if only temporarily.
   Even when the p.t.o. is disengaged, it is safer to stop the tractor engine before working on the machine.
2. Do not work under implements held in the raised position by the tractor hydraulics.
3. Never leave the driving seat while the outfit is in motion.
4. Always attach trailed implements to the proper drawbar and never to a high position on the rear of the tractor.
5. Always use a proper draw pin with a locking device.
6. Never leave *self-lift* implements in the raised position.
7. Manhandling of rollers when attaching them is a common cause of stomach strains.
8. Allow adequate headlands especially when using trailed implements near banks, rivers and ditches.

9. Ensure the tractor is jacked securely when altering wheels for row-crop work.

10. Avoid the temptation to clear toothed harrows by lifting them manually, when the outfit is in motion, and never leave such harrows inverted in the field after use.

3 MANURE AND FERTILISER DISTRIBUTION EQUIPMENT

The value of farmyard manure—its benefits to soil structure and in the supply of plant nutrients—are well known, although its precise behaviour and effects are still the subject of research.

To some extent, the actual benefit derived from farmyard manure in the production of particular crops depends on the timing and method of its introduction to the soil, as well as on the quantity and nature of the manure applied. These factors have to be taken into account when deciding on a suitable manure-handling system for a particular farm. Other factors are cost of equipment, the size and pattern of the farm, the general layout of yards and buildings, and the labour available. Even the type of soil is relevant, as it may not be capable of supporting the weight of manure spreaders during the winter months without damage to its structure.

Construction, operation and maintenance of tractor-mounted loaders

Before the introduction of tractor hydraulics, few tractor loaders were seen, because of the need for a cumbersome superstructure to actuate the lift arms. Now, with the aid of external hydraulic

rams, the problem has been simplified, and a wide range of
loaders is marketed. Most of these are of the forward-action type,
and are referred to as *front mounted*, although their supporting
framework is often positioned towards the rear of the tractor (Fig
90). Front-mounted loaders are more versatile than rear-mounted

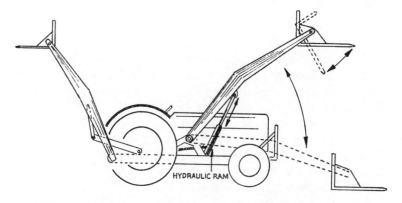

90 Front-mounted and rear-mounted hydraulic loaders

ones, and give the driver better visibility. Rear-mounted loaders,
however, improve wheel grip because the weight of the loader
and its pay-load are concentrated on the rear wheels of the
tractor. Also the rear tyres of a tractor are better able to carry the
additional load than are the front tyres. Some rear loaders are
adapted to provide a side-slewing effect; this is made possible by
the greater stability of the loader at the rear of the tractor.

The various modes of attachment call for varying degrees of
skill in fitting loaders to tractors, and in some cases slight modifications have to be made to the tractor itself. A wide range of
attachments is usually available in addition to the manure fork,
shovel and pallet fork shown in Fig 91; these include bale loaders,
crop-collection forks, ditching attachments and hedge trimmers.

The whole assembly consists of a robust superstructure incorporating two hydraulic rams, one on either side of the tractor.
The actuating rams are connected to the external tappings of the
tractor hydraulic system by flexible hoses (see Plate 3).

SHOVEL   MANURE FORK   PALLET FORK

91 Three typical loader attachments

The *reach* of a loader varies with the design, and influences its maximum operating height. Alternative loader arms are available from some makers to suit specific requirements. Some loaders have alternative pivot points for the lift arms, permitting an adjustable maximum height. This increased height sometimes results in a slight reduction in the loader's *tear-away* effort, i.e. the effort required to extract the loaded fork from the manure heap. This effort can be considerable when the manure is firmly consolidated, especially if it contains long, unrotted straw. Loaders differ in their reach from 8 to 12 ft. The specified lifting capacities of various makes range between 700 lb and 1,500 lb, and the tear-away effort usually exceeds the lift capacity by 30% to 40%.

The efficiency and operational speed of hydraulic loaders depend largely upon the characteristics of the tractor hydraulic systems which power them. The introduction of 'live' hydraulics on tractors (i.e. hydraulic systems which operate independently of the transmission clutch) has substantially simplified and quickened loader operation. Tractors with an independent oil reservoir for the hydraulic system have an advantage, particularly in colder weather, over those which utilise the transmission oil. The lighter grade of oil used in independent hydraulic systems flows more freely, and this speeds up the loader's reaction to the hydraulic controls.

The variety of hydraulic control systems available make it essential for tractor drivers to familiarise themselves with the one in use before attempting to operate the loader. Incorrect operation can cause prolonged periods of overload on the hydraulic system.

The mechanism for controlling the tilting and relatching of the fork itself varies in detail with different makes of loader. In some cases, the action relies upon correct balance and swing of the fork. In others, a positive control is provided, which is an advantage when handling slow discharging material that tends to hang on the fork and unbalance it. With some forks, the material is discharged by a pusher plate which slides along the tines. Sometimes, parallel links are provided to keep the fork or other attachment level throughout the lifting range.

ATTACHMENT OF LOADER TO TRACTOR AND PREPARATION FOR WORK

The attachment of the loader and its superstructure can be complicated although manufacturers usually supply explicit instructions. Strict adherence to the recommended procedure can save much time and energy, and although the initial attachment is usually a two-man job, the subsequent removal and re-attachment of the loader arms is relatively easy.

Sometimes parts of the assembly conceal vital maintenance points on the tractor, but in this event, the loader manufacturer supplies all the modifications necessary to reduce this inconvenience. The hydraulic system unions must be clean and in good condition before the initial connection. On some makes of tractor, a valve permits isolation of the external hydraulic circuit from the three-point linkage circuit.

Most tractors are now fitted with heavy-duty front tyres capable of carrying the extra weight imposed upon the front of the tractor; it is advisable always to fit such tyres. It is inadvisable to inflate standard tyres to excessive pressures, although a slight increase of about 2 or 3 lb/in^2 is tolerable and often recommended. Ballasting of the rear tyres, or the attachment of ballast weights to the rear of the tractor, is sometimes recommended when front loaders are fitted, in order to improve rear tyre adhesion and to help stabilise the tractor. To improve the stability of the tractor still further, its wheels may be set out wider than their standard position. This is particularly desirable when working on a hillside or on uneven land. A guard may be fitted at the front of the tractor to protect the radiator.

OPERATION

A tractor-mounted manure loader is not as easy to operate as it may seem. The novice invariably lunges into the heap with great optimism, but soon discovers that the technique of mechanical handling only comes with practice. Various techniques have to be employed, according to the type of work on hand. Manure may have to be lifted from a stockyard floor where it has become firmly consolidated, from a heap in a yard with restricted manœuvring space, or from a field heap where the ground may be soft or uneven. The way in which a manure heap has been built, and its degree of decomposition, will also influence operating technique and loading procedure. The following points will be helpful to the beginner:

1. When extracting manure from a heap, the upper layers should be tackled first. The best depth of *dig* will depend upon the state of the material. If it is well rotted, less tear-away power is needed, and substantial fork loads can be taken. Once an operator has acquired the *feel* of the loader, he will find it an advantage to raise the fork gradually as the tractor moves into the heap.

2. When tackling manure which has been firmly trodden down by stock, it is sometimes difficult to penetrate the upper layer of un-rotted material. In this case, it is usually possible to skim off the surface by setting the fork a few inches into the material and driving the tractor forward.

3. Deep, well-rotted manure can sometimes be loaded more easily if vertical cuts are first made in the heap with a hay knife.

4. Take special care when picking up manure from soft or uneven ground, as the fork tines tend to dig in. It is advisable, as far as possible, to site manure heaps on a flat concrete surface.

5. Always load the fork evenly across its width, to prevent twisting of the loader frame by uneven stress. Side-thrust on the loader should be avoided by approaching the heap on a straight course.

6. When loading a spreader, start at the end farthest from the spreading mechanism. The height of the load should not exceed the top shredding beater.

Properly used, there is no reason why a loader should cause any undue wear and tear on a tractor, but carelessness may produce excessive strains. Driving violently into a consolidated manure heap causes severe shock to the loader and its supporting frame. Careless shunting and harsh use of independent brakes, especially on rough yard bottoms, can severely damage the tractor tyres, and put unnecessary strains on the steering, transmission and braking systems.

MAINTENANCE

It is important to grease the pivot points on the loader regularly to prevent the entry of manure with its corrosive effects. Occasional cleaning also helps to control corrosion on the loader as a whole. A regular check should be made to see that the loader's supporting structure is secure on the tractor, as the loss of one securing bolt may result in serious damage to part of the tractor engine or transmission casing. If any of the manure fork tines become bent they should be straightened. If the loader is not in regular use, it is usual to remove it from the tractor between periods of service but the supporting frame can usually remain without causing inconvenience. If the hydraulic rams are not removed with the loader, they should be secured to the tractor in their contracted position and protected from corrosion.

Construction, operation and maintenance of trailer-type manure spreaders

Trailer-type manure spreaders are commonly used in conjunction with tractor loaders. There is such a wide range of makes on the market that prospective purchasers find it difficult to make a choice, for although many spreaders may be similar in appearance they possess important variations in the design of their components. Both forward-delivery and rear-delivery versions are available, although the latter is much more popular at the present time and the following description relates to its various components (see Fig 92).

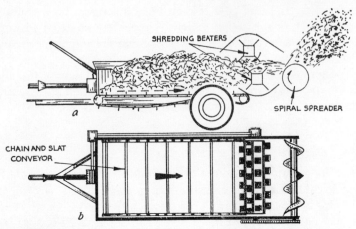

92 Trailer manure spreader: *a* Elevation, *b* Plan of spreader with
flail type shredders

Chassis and body

The chassis and body are generally made in one unit, forming a
robust two-wheel trailer with front and side panels. At the rear of
the body, two side 'cheeks' are fitted to accommodate the spreading mechanism. The construction is either all steel, or steel
framework with wooden bottom and sides. The steelwork is
usually treated against corrosion, and the woodwork is pickled in
creosote to check rotting. The capacity of the body may be anything between 60 and 120 bushels, i.e. $1\frac{1}{2}$ and 3 tons approximately. Although the larger-capacity spreader reduces the number of
journeys needed to transport a season's manure accumulation
from heap to field, the smaller machine is convenient for hauling
the daily collection direct from yard to field, especially where wet-
lying land might suffer from the weight of a larger spreader.

The wheels of the spreader are always set well back to transfer
some of the loaded weight on to the rear of the tractor to improve
wheel adhesion. As a result, a screw-type parking jack or a ring-
type mechanical attachment is essential. Large section tyres are
used to improve flotation on soft ground, and their large diameter
ensures good ground clearance under the axle.

1. Single-furrow plough

2. Three-furrow reversible plough

3. Hydraulic loader

4. Precision seeder units

Chain and slat conveyor (apron)

The conveyor moves the manure towards the spreading mechanism when the latter is in operation. Its movement may be continuous or intermittent. In either case, its speed can be regulated to control the rate of manure discharge.

A few spreaders have been marketed with a rubber conveyor as an alternative to the chain and slat type.

Shredding beaters

Machines may have one, two, or even three, shredding beaters; they are cylindrical assemblies made up of rake-bars, each carrying a number of tines. If there are more than one, they may be designed to rotate at slightly differing speeds. The function of the beaters is to tear apart lumps of manure, to ensure a constant and even delivery to the spiral spreader.

Spiral spreader

The spiral spreader is a left- and right-hand auger which revolves at high speed and scatters the manure evenly over a wide strip of land. On machines having only one shredding beater, the spiral usually has blades or tines attached to it to lacerate and chop the material as it is distributed.

System of drive

Both power-driven and ground-wheel-driven machines are available. The tyres of ground-driven machines have land-grip treads, capable of providing all the power necessary when traction conditions are good. Ground-wheel drive eliminates the extra work of coupling and uncoupling the power-drive shaft when one tractor is being used for both loading and spreading. Power-driven machines, on the other hand, are preferable when spreading in bad traction conditions and where long, unrotted or lumpy

E

manure has to be handled. Also, a greater range of operating speeds is available by driving the tractor in various gears.

The beaters and spreader are driven by chains, and the intermittent conveyor motion, where required, is achieved by means of a rachet and pawl mechanism. Slip-clutches, designed to prevent overload of working parts, are fitted to most machines either as one master clutch or, as is more desirable, individually on each principal shaft.

Convertability

Some makes of spreader are designed so that the distributing mechanism can be quickly removed from the rear of the body to convert the loader into a general-purpose or moving-floor trailer. On other makes, the rear spreading mechanism can be replaced by attachments for spreading artificial fertiliser and lime.

Rear-delivery spreader attachments

Fig 93 illustrates an attachment designed for quick fitment to, and removal from, a trailer. Drive for the shredder and distribution mechanism is taken from the tractor p.t.o. The manure is moved rearwards to the spreading attachment by a conveyor or scraper-blade. Speed of the spreading mechanism can be varied to regulate the rate of application.

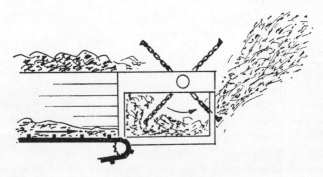

93 Manure spreader attachment (chain flail type)

ROTARY SPREADER

One successful departure from the conventional type of trailer spreader is illustrated in Fig 94. It has a watertight cylindrical container, the contents of which are discharged by revolving chain

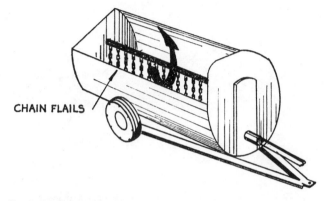

CHAIN FLAILS

94 Rotary manure spreader

flails carried on a common spindle and powered by the tractor p.t.o.

This revolutionary design has a number of advantages: it can handle both solid and liquid manure; it achieves a high degree of pulverisation of the solid manure; it gives a consistent spread; its simplicity of design makes for reliability and the minimum of maintenance requirements.

Construction and operation of heap spreaders

A heap spreader is usually a mounted or semi-mounted power-driven machine employing the same basic principles as the spreading mechanism of a trailer spreader. Although few are used today, they are relatively cheap and successfully handle the manure-spreading problems of small farms.

The manure must first be moved from the yard, buildings or heap, to the field, and deposited in small heaps at regular intervals in rows 12 to 15 ft apart. This may be done with a tipping trailer. The spreading outfit is then driven down the rows straddling the heaps; as with trailer-type machines, a shredding beater disintegrates the material and a spiral spreader distributes it (Fig 95).

95 Field heap manure spreader

Reducing the ground speed by means of the tractor gear box gives greater pulverisation of the manure, and a second 'pass' is sometimes made to improve further the distribution.

OPERATION

Trailer spreaders usually have one lever to engage the drive to the shredding beaters and spiral spreader, and a second lever to engage the drive and adjust the speed of the conveyor. A further speed range for the conveyor can often be obtained by altering the effective stroke of its driving ratchet pawl. Various settings provide application rates ranging from 5 to 25 tons per acre.

It has been pointed out previously that loading should start at the far end from the spreading machanism, and that the height of the load should not exceed the top shredding beater. These precautions are necessary to ensure a smooth feed and to avoid undue strain on the mechanism. The drive to the beaters and spreaders should always be engaged before that to the conveyor. Speed of travel, ground conditions permitting, can be up to 5 miles per hour.

MAINTENANCE

Spreaders are available in a wide range of qualities (and prices). Quality is very important with this equipment, which comes into continual contact with corrosive substances. Much unnecessary damage occurs to manure-handling machinery when it is out of use. Even a spreader in more or less constant use benefits from an occasional hosing down. This is particularly important in frosty weather when frozen manure tends to lock the working parts.

If a high-pressure water jet or steam cleaner is used on machines, water must not be directed into the bearings. When most of the water has drained off, a coating of water-displacing anti-rust compound will give at least temporary protection. After the beater spindles are cleaned, their bearings should be examined for any straw, baling twine, or other fibrous matter entwined around the shafts. Lubrication should follow the manufacturer's recommendations exactly, as different bearings require different treatments. In some cases, sealed bearings are used, and these may or may not require periodic recharging with grease.

The periodic dismantling, cleaning and reassembling of slip clutches is important, as serious damage can result if these fail in an emergency. Obstruction is commonly caused in the machine by bricks or stones picked up by the loader when filling the spreader. Chain tensions should be checked, and set so that $\frac{1}{2}$ in deflection is possible for each foot between sprocket centres. Sprockets should be replaced as soon as wear on their teeth is apparent, or rapid chain wear will result.

Tyre pressures also are important, especially where rough tracks and yards have to be negotiated. If a tyre on a ground-driven machine is replaced, it must be fitted to rotate in its effective direction, i.e. the direction *opposite* to that indicated by the arrow on the tyre. Unlike the rear tyre of a tractor, it is driven by contact with the ground.

Construction and operation of liquid-manure distribution equipment

The wastage of valuable plant nutrients through stockyard

drainage systems has always caused concern, but its collection and distribution is problematical. It is only comparatively recently that constructive thought has been given to methods of transferring the liquid from farmstead to field. Additional stimulus has arisen from two further factors:

1. Legislative pressures designed to reduce the contamination of water courses by farm effluent.

2. The increasing number of farms being equipped with irrigation systems through which the manure can be distributed.

LIQUID-MANURE TANKERS

Tankers for liquid manure are available as large-capacity models ranging from 200 to 300 gallons (Fig 96), or small models designed for mounting on the tractor three-point linkage (Fig 97). The tank is generally galvanised or coated internally with an anti-corrosive substance. The liquid manure is transferred from its collecting sump into the mobile tank either by suction achieved with a vacuum pump or by a centrifugal pump, both of which are available for tractor p.t.o. drive. The liquid is subsequently discharged from the tank on to a splash plate or rotating distributing disc with the outfit in motion. The timing of the operation and the rate of application must be right if full benefit is to be obtained from its manurial value. Care is necessary also to avoid scorching crops or grassland by overdosing with nitrogen-rich liquid.

ORGANIC IRRIGATION

The layout of any organic irrigation system depends on the size, siting and elevation of the buildings and yards included in the scheme. The essentials are a collecting tank, a pump and agitating mechanism, distribution lines and rain guns. An outline of each of these follows:

Collecting tank

The size of the tank depends upon such factors as the number of stock accommodated in the buildings or yards, whether or not all

solid manure is to be washed into the pit, the area involved, and
the frequency of washing down.

Road and roof drainage should not be collected in the pit, or it
might flood after heavy rain. The pit should be covered and fenced

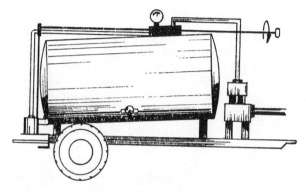

96 Liquid-manure tanker (trailed type)

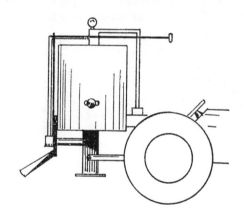

97 Liquid-manure tanker (mounted type)

off, and a remote overflow outlet should be provided so that the
pit itself does not flood.

If solid manure is to be conveyed into the pit, it must be washed
in with ample water. If the contents of the pit are insufficiently

diluted, pumping may be difficult. Long straw, silage and other debris such as pieces of wood, stones, twigs, etc. should be kept out as they could damage the pump or block the system.

Pump and agitating mechanism

A piston or centrifugal pump may be used to transmit the slurry from the tank to the field. Greater efficiency can be expected of the piston-type pump, which can operate at higher pressures, so reducing blockages and increasing the velocity of the liquid at the rain-gun nozzles. The higher pressure capacity of the piston pump is essential for long deliveries and where the rain guns are located in elevated positions. Centrifugal pumps are not so efficient, but are cheaper and quite commonly used where the system is not too demanding. Sometimes the pumps are fitted with choppers to cut long straw passing through the system. Pumps may be driven either electrically or by the tractor p.t.o.

In some systems, a paddle-type agitator is used to stir up the slurry in the tank before and during pumping; in others, the agitator is built into the pump itself.

Pipes and rain guns

Standard irrigation pipes are used, although liquid manure needs some modification of the line valves. The effective coverage area of rain guns may range from a quarter to three-quarters of an acre, depending upon the pressure available and the size of nozzle used. The nozzle must be large enough not to cause blockages, and rubber nozzles are available which are capable of discharging solid particles which would block the standard metal type.

Construction and operation of fertiliser distributors

In spite of the benefits derived from applications of farmyard manure to the soil, few crops produce their maximum yield without specific supplementary nutrients in the form of 'artificial' fertilisers, i.e. activating elements chemically synthesised into a form in which they can be absorbed into the soil. These fertilisers may be applied to the soil immediately before a crop is sown, or later, when the crop has germinated and become established, in an operation commonly known as *top-dressing*. A third method is to apply the fertiliser simultaneously with the seed by means of a combine drill. This drill is discussed in Chapter 4. One example of the actual dispensing mechanism commonly used in combine grain drills is shown in Fig 113.

Some fertiliser distributors may also be used for the application of lime, gypsum, herbicide and insecticide dusts, and even small seeds.

Considerable research has gone into the design of fertiliser distributors, and several types have evolved which give very satisfactory results. A major problem is to achieve consistent and even discharge of fertilisers which are not always in free-running condition because of their susceptibility to physical change particularly in a damp atmosphere. Some fertilisers become lumpy, others become sticky, causing them to pack or *bridge* in distributor hoppers. Fertiliser deterioration is sometimes the result of long periods of storage in unsuitable premises. Few fertiliser distributors can handle lumpy material satisfactorily, but some are more tolerant than others. Certain fertilisers are available in granulated form; this reduces their tendency to cake and helps them flow more freely. Fortunately, a little unevenness in fertiliser distribution is not generally very serious.

The various machines available differ in the type of feed mechanism employed. For this reason, distributors are usually classified by this feature. The following comparative study of the alternatives is limited to the more widely used types.

Reciprocating plate distributor

Fig 98 illustrates the well-established reciprocating-plate type of distributor. The fertiliser discharge is achieved by the action of two slotted plates, positioned one above and one below a fixed aperture plate at the bottom of the hopper. The slotted plates slide with a reciprocating action along the upper and lower faces of the aperture plate and filter the fertiliser through. The application rate may be regulated either by varying the 'throw' of the reciprocating plates, or by using other plates having different-sized slots.

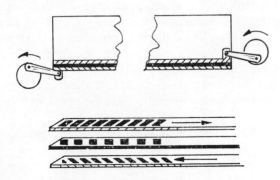

98 Reciprocating plate fertiliser distributor

Rotating plate and flicker distributor

The rotating plate and flicker type of feed is also used extensively. Its working principles can be seen in Fig 99. The concave discs

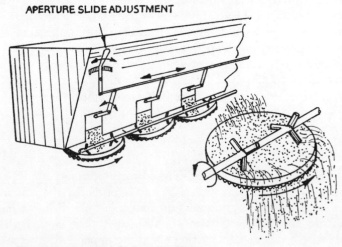

APERTURE SLIDE ADJUSTMENT

99 Rotating plate and flicker distributor

rotate continuously, receiving a constant flow of fertiliser from the hopper. The fertiliser is flicked over the edge of the discs by small fingers rotating in the vertical plane just above the surface of the discs.

The rate of application is regulated by alternative settings of the hopper shutters and by varying the rotational speeds of either or both of the moving components. The distributor feed can be quickly cut off by a single control lever at the bout ends. Both trailed and mounted versions are available, and the latter type sometimes has power drive to the feed mechanism. This increases the range of sowing rates obtainable. When using power-driven models, it is important to establish, and maintain, the appropriate ground speed and gear ratio on the tractor. Tractors with multi-speed gear boxes giving a widely variable p.t.o./ground-speed ratio are a great asset when using this type of distributor.

Spinning-disc distributor

The spinning-disc type of feed mechanism influences the overall design of the distributor as can be seen from Fig 100. The cone-

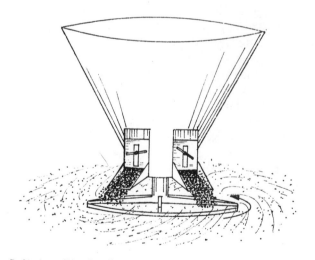

100 Spinning-disc distributor

shaped hopper, which may have a capacity of 3 to 10 cwt, usually incorporates an agitator to keep the material flowing. The agitator in the base of the hopper is usually capable of breaking up large lumps of fertiliser as long as they are soft. Both single and double disc versions are available.

Depending on whether the machine is attached to the three-point linkage or trailed, the spinning disc (or discs) which scatter the fertiliser may be driven by the tractor p.t.o., by land wheels, or by means of a hydraulic motor powered by the tractor hydraulic system. The fertiliser delivery is regulated by an adjustable shutter. Two valuable features of this type of distributor are its narrow transport width and its wide operational span. The spreading width may range from 15 to 50 ft according to the speed of the rotating disc(s), and in some cases the adjustment of the discharge shute between the hopper and the disc(s). The type and condition of the fertiliser affects the range and evenness of distribution.

When attaching the mounted-type machine, the top link should be adjusted to set the hopper vertical. About 5 or 6 miles per hour is a good speed in most conditions, but a rough surface, or the stage of a crop's growth when top-dressing, may sometimes dictate lower ground speeds. Careless driving when top-dressing can damage the crop. The simple construction of this type of machine facilitates cleaning and corrosion control measures; some makes have plastic or glass-fibre components to reduce corrosion damage. A criticism of the spinner type of distributor is its tendency to throw fertiliser on to the rear of the tractor, causing corrosion of the external equipment. Damage can be avoided by cleaning the rear of the tractor off when the fertiliser distribution is finished. Unnecessary stops with the distributor should be avoided, as they cause fertiliser to accumulate on the spinner disc(s), with a subsequent heavy discharge when the machine re-starts. The width of spread makes it difficult to match the bouts, and one company has devised a novel system whereby marker poles are positioned along the headland at intervals in accordance with the machine's effective spreading width, and a device on the front of the tractor collects the appropriate pole at the end of each bout. Even when using markers, bout matching is almost impossible in windy weather. If work has to proceed in such conditions, a lower disc speed and narrower bout is preferable.

Impeller and pendulum-spout distributor

The general layout of the impeller and pendulum-spout type, as shown in Fig 101, is similar to that of the spinner type, although the actual distributing mechanism consists of an impeller and a

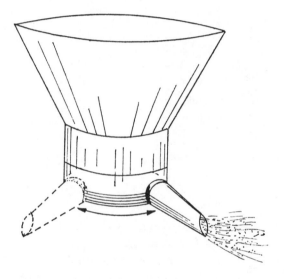

101 Pendulum-spout distributor

swinging-pendulum spout. This method of discharge gives better control over the effective operational width. It prevents fertiliser being thrown on to the tractor, and is claimed to be most effective for the sowing of fine seeds, particularly if the latter are mixed with a carrier.

Radial-spout distributor

The general layout of a radial-spout type of distributor is similar to that described above, except that the fertiliser is discharged through radial spouts from an internal impeller (Fig 102).

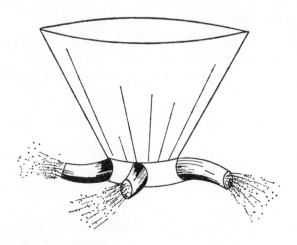

102 Radial-spout distributor

Liquid fertiliser applicator

This is a recent development which is proving most effective. Sharp knife-like tines make possible the injection of liquid plant nutrients directly into the soil. This eliminates the problems sometimes encountered with solid fertilisers and supplies the nutrients in a more assimilable form, thus giving a quicker response to the treatment.

Moving-floor distributor

Recently, interest has grown in large capacity, moving-floor-type distributors. As can be seen from Fig 103 such a distributor consists of a trailer constructed with a vee-shaped trough, at the bottom of which is an auger or slat-type conveyor to carry the material to the rear. At this point, the contents are dropped on to one or two rotating spinners for distribution. The spinners may be powered by the tractor p.t.o., or by the tractor hydraulic system through the medium of a hydraulic motor.

The rate of application is usually regulated by an adjustable

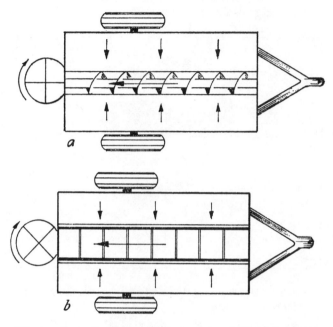

103 Moving floor distributors: *a* Auger type, *b* Slatted conveyor type

rear door or slide, and by altering the speed of the auger or conveyor.

Two- to three-ton capacities are usual, and the spreading width generally ranges from 30 to 50 ft.

Fertiliser placement attachments

Some special-purpose distributors have been designed to place fertiliser accurately in relation to growing crops. Such machines have mostly been superceded by *placement* attachments, which can be fitted to standard distributors or other machines such as potato ridgers, row-crop toolbars, precision drills, etc. The object is to place the fertiliser close to the seeds or plants yet not actually in contact with them as this might cause damage to the seedlings or tubers. Fig 104 gives some examples of placements.

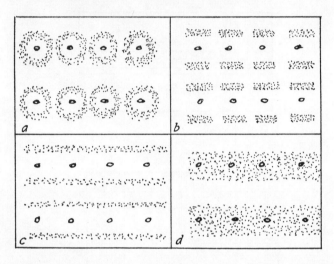

104 Fertiliser placements: *a* Ring, *b* Intermittent band, *c* Adjacent strip, *d* Continuous band

Maintenance

Many distributor manufacturers now fit sealed bearings and stainless steel, glass fibre and plastics components which are impervious to corrosion damage. Components in contact with fertilisers are made accessible so that they may be quickly removed for cleaning. Reference should be made to the manufacturer's instructions for the lubrication and maintenance requirements of particular machines. The importance of dismantling, thorough cleaning of all working parts, and the application of a rust preventive before storing away cannot be overstressed. The rust preventive should be removed before work begins. Chains are used on many fertiliser distributors, and these prove to be one of the worst victims of corrosion, particularly when they are enclosed in a guard and out of sight (unless, of course, they are in an oil-tight casing). It is advisable to remove these, clean them and store them in a lubricant until again required. Even when a machine has been treated in this way before storage, it is advisable after re-assembly and before work, to test the mechanism manually before engaging either the land wheel or power drive.

Safety precautions

Farmyard manure-handling equipment presents a number of potentially dangerous situations, although the machines are not particularly hazardous; some accidents do occur each year as a result of thoughtlessness often caused by the mis-handling of tractor-mounted loaders. Operators have also become victims of the shredding and distributing mechanism of trailer-type spreaders. The following precautions should be taken.

1. Operate tractors and loaders with care, especially when other people are in the vicinity.

2. Do not drive carelessly around farmyards and buildings with the loader lowered.

3. Take care when driving under low electricity cables with the loader raised.

4. Lower the attachment to the ground before carrying out tractor maintenance, and always when leaving the outfit—if only for a short time.

5. Never interfere with the ram-hose connections when the loader is raised or the system otherwise under pressure.

6. Remember that the tractor is less stable when the loader is raised. This is particularly important when working on hillsides or on very uneven land.

7. Shield the tines of manure forks before travelling on the highway.

8. Ensure that no one is near the distributing mechanism of trailer spreaders before engaging the drive.

9. Never stand in the spreader with the tractor p.t.o. running.

10. Never attempt to clear blockages, lubricate or adjust the mechanism of a spreader with the latter running.

4 DRILLS, SEEDERS AND POTATO PLANTERS

Before the introduction of the corn drill, seed was scattered by hand and then harrowed in. This broadcasting of the seed was a laborious business, and considerable skill was required to distribute the seed evenly. When the grain drill was first introduced, its sowing accuracy and its ability to place the seed in the soil were quickly recognised, and the drill soon became, and remains, a most important item in the farmer's complement of machines. Although they are produced by many manufacturers, drills are of a fairly uniform design. However, there have recently been some departures from the traditional patterns and these are discussed later in this chapter.

Root-seed drills and potato planters have also reduced the problems of crop production by eliminating much of the manual labour traditionally entailed in the growing of root crops. So far as these machines are concerned, however, there is a wide variety of designs on the market, and the features of some of these are also outlined.

Construction and working principles of grain drills

The conventional drill is not a complicated machine, as can be seen from Fig 105. A few minutes' inspection of such a machine should not leave many mysteries uncovered.

Basically, it consists of a main frame supporting a hopper which incorporates the seed-metering mechanism on its underside. As with most other implements, mounted and trailed versions are

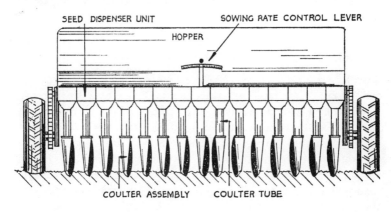

105 Grain drill (rear view)

available, although in this case the mounted type has no great advantage over the trailed type. Most of those designed for three-point linkage attachment still retain land wheels for accurate depth control. The dispenser units of the seed-metering mechanism, which are generally driven by the land wheels of the drill, transfer the seed from the hopper into the flexible coulter tubes beneath. From these, the seed is supplied to the coulters which open up narrow grooves (or *drills*) in the soil to receive it.

HOPPERS

The hopper, which may be of wood or steel, has vertical partitions inside to keep the seed spread out over its full width. This is parti-cularly important when drilling across hillside land, when the seed tends to run to the lower end of the hopper and leave some of the seed-metering units at the other end starved. Specially

129

designed hoppers are necessary for cup-feed drills. These embody a separate seed reservoir which ensures a steady supply of seed to the cups. On some force-feed drills, an agitator is incorporated in the bottom of the hopper. This is simply a spindle with projecting tines, running the full width of the hopper. It rotates slowly to keep the seed loose and free-flowing immediately above the dispenser units.

SEED-DISPENSING MECHANISMS

The function of the seed-dispensing mechanism is to transfer seed from the hopper into the coulter tubes at a pre-set rate synchronised with the ground speed of the drill. The mechanism is driven by the land wheels of the drill through a train of drive employing either gears or chains or both. The sowing rate must be accurately adjustable in spite of the widely differing sizes of the seeds commonly sown. The difficulty this presents accounts for the many different types of seed-dispensing mechanism available, three of which are described below.

Cup feed

The cup-feed mechanism has been in use for many years, and is still favoured by many farmers for sowing root seed. The seed is dispensed by several rotating discs—usually one for each pair of coulters. Each disc has a series of cups on either face which, as the discs rotate, collect seed from a shallow reservoir fed from the main hopper, and discharge it into the top of the coulter tube (Fig 106).

The efficiency of this feed mechanism is dependent upon a steady supply of seed from the main hopper to the compartment housing the rotating cups. For this reason, the hopper must be kept upright, and adjustment must be made when working either uphill or downhill. Excessive ground speed must also be avoided to prevent seed from being shaken out of the cups. This has become a disadvantage since tractors replaced horse power.

Advantages are its gentle action, which is unlikely to damage the seed by abrasion or crushing, and its ability to sow with acceptable accuracy, anything from fine grass seeds to the largest of cereal and root seeds.

The sowing rate may be regulated by altering the speed of the rotating cups—usually by means of a variable ratio gear or chain drive, or by the use of alternative-sized cups. In the latter case,

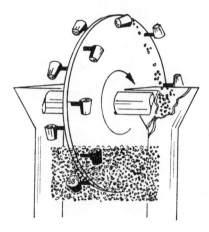

106 Creep feed mechanism

the whole seed *barrel*, i.e. the complete array of discs and cups and the spindle on which they are assembled, is either changed altogether, or reversed, i.e. changed end for end to bring into operation the other sides of double-faced cups which will be either smaller or larger as shown. Changing the seed barrel is also necessary to adapt the drill for different sized seeds.

External force feed

The external force-feed type of seed-metering mechanism was introduced to overcome the drawbacks of the cup-feed type. *External* implies that the periphery of the metering wheel propels the seed from the hopper into the coulter tube, and *force* indicates that this is performed with a more positive action. The individual units—one for each coulter—are driven by a common shaft from the land wheels of the drill. One commonly-used type of metering wheel is shown in Fig 107. This takes the form of a fluted roller which propels the seed around the inner face of a concave plate before discharging it into the coulter tube. The sowing rate is usually adjusted on the fluted roller type by sliding the rollers

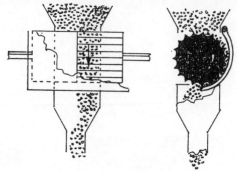

107 External force feed mechanism (front and side elevation)

sideways relative to the hopper orifices, and using a greater or lesser width of the rotating element. Varying the speed of the rollers also can affect the sowing rate.

Another type of wheel, in current use, has a series of tooth-like projections, and runs at a higher speed than the fluted roller. It is sometimes referred to as the *speed-feed* type (Fig 108). With

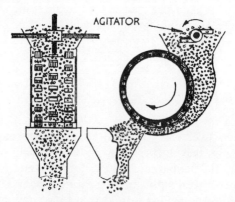

108 External speed feed mechanism

both these mechanisms, accurate metering is attainable over a wide range of seeds.

The metering wheels of the speed-feed mechanism are claimed to give a versatile and even delivery rate without damaging the seed by constriction of the *run*, as sometimes occurs with other types.

Sowing rate is regulated on this type of feed by varying the speed of the toothed wheel.

Internal force feed

The internal force-feed mechanism is widely used, but although it is satisfactory for drilling the more common cereals, it is not as versatile as the types described above. It may sometimes be used for sowing very large or small seeds, but its accuracy in handling these extremes does not equal that of the external type.

It has a revolving wheel with a wide rim which gives a shallow cylindrical or concave effect on both faces. These faces are serrated —finer on one side than on the other—and choice of face provides a means of adjustment for seeds of differing sizes (Fig 109). The serrations impel the seed through the restricted space between the wheel and its housing.

Sowing-rate adjustment is made by varying the speed of the

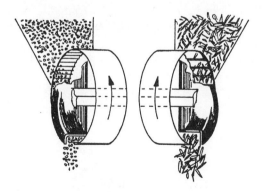

109 Internal force feed mechanism: *a Fine* side, *b Coarse* side of dispenser unit

wheel. This usually involves altering the position of a pinion which meshes with a crown wheel having different series of teeth at various radii. When the appropriate side of the dual-faced metering wheel has been decided for a particular seed, a hinged or transferable cover-plate blanks off the inoperative side of the wheel. The cover-plates are usually marked *fine* and *coarse* to indicate the side of the metering wheel exposed. The fine and coarse sides are used for small and large seed respectively.

Centrifugal feed

This is a major departure in seed dispensing mechanism. The seed is delivered from a central revolving cone by centrifugal action to the coulter tubes which are positioned around a radial manifold. The revolving cone is driven by the land wheels, and is therefore related to the ground speed. The efficiency of this method of distribution is such that a wide range of drilling speeds may be used— the rotating cone itself running at speeds of anything from 300 to 1,200 rev/min. The centrifugal force created by such speeds ensures positive seed delivery. The machine is able to handle a wide range of seeds.

COULTER TUBES

Coulter tubes, which convey the seed from the metering units down to the coulters, have to be flexible to accommodate the movement of the coulters over the ground. Spiral steel tubes have been extensively used for some years, but are now rendered obsolescent by rubber, rubberised canvas and plastics types, particularly on combine drills where they suffer seriously from corrosion.

COULTERS

Sometimes referred to as *furrow openers*, the coulters on corn drills are generally spaced at 7-in intervals across the width of the drill. There is a recent trend towards closer spacings, and on some drills the spacing is variable.

Coulters cut a narrow groove in the soil to take the seed at the right depth and in a continuous line. The number of coulters on any particular drill obviously depends upon the width of the drill, and this commonly ranges from 8 to 12 ft. Various types of coulter are in current use, although the most common are the disc type, of which there are two versions—single disc and double disc, the hoe type and the Suffolk type.

Single-disc coulters

The single-disc coulters are concave, and carried on spigot spindles (Fig 110). They are set at a slight angle to the direction of travel to give their concave faces a slight forward inclination. This helps their entry into the soil, and determines the width of the resultant groove. The cutting action of the discs helps to improve the condition of the seedbed, particularly if there is a shortage of

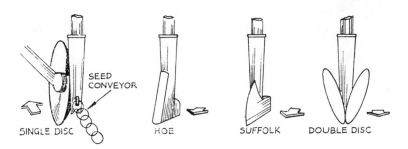

110 Types of coulter

tilth. On stony land, the disc coulter may be at a slight disadvantage compared with the hoe type described below. Disc coulters also tend to block when drilling in moist soil conditions, in spite of the provision of scrapers designed to check this problem. Most farmers and tractor drivers are well aware that drilling should not proceed under such circumstances, but emergencies often arise and drilling sometimes has to go on even though soil conditions are not ideal.

135

Double-disc coulters

The discs of a double-disc coulter may be concave or flat. Each pair is arranged in a vee formation with their leading edges close together (Fig 110). This eliminates the slight side-thrust suffered by single discs due to their slight angular disposition, together with the misalignment of the drills which sometimes results.

Higher ground speeds are possible with the twin discs, because their shielding action prevents the seed rolling after delivery from the coulter tubes. Also, the movement of the soil on either side of the drill can have a valuable cleaning effect on stale seedbeds. Some of the points made above in relation to the single-disc coulter apply also to the double-disc arrangement.

Hoe-type coulters

Fig 110 also illustrates a hoe-type coulter. It has no moving parts, which simplifies maintenance. In some cases, its leading edge has a bevel which provides a slight cutting action; in others, the leading edge is curved and merely parts the soil to form a narrow seed channel. The hoe coulter is often favoured on heavy and wet soils —being free of moving parts which would tend to become blocked in such conditions. It is also sometimes preferred on stony, light and abrasive soils because of its mechanical tolerance of such conditions.

Suffolk-type coulters

These are similar to the hoe type but have a sharp leading edge to minimise soil disturbance (Fig 110).

COULTER FRAMES

Each coulter is carried on an individual frame which is pivoted at its forward end so that it can rise and fall with the ground contours. An adjustable tension spring prevents bouncing, and controls the depth of coulter penetration (Fig 111). As a rule, two lengths of frame are used, and these alternate long and short across the width of the drill frame so that the coulters are staggered. This allows any stones, clods and trash to filter between the coulters without blocking.

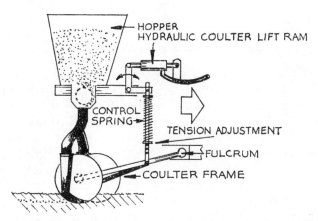

111 Side view of drill showing coulter frame and control spring

COULTER LIFT MECHANISM

A lift mechanism is necessary to lower and raise the coulters into and out of work during the drilling operation. On some drills, a lever is provided for manual operation by a second man riding on a platform at the rear of the drill. Various retaining positions on the lever ratchet or quadrant determine the coulter depth. On other drills, a mechanical coulter lift is fitted and the depth setting is separate. The lift unit is commonly powered by the land wheels of the drill, and the trip mechanism which actuates it is controlled by the tractor driver by means of a cord. A more recent innovation involves the use of a remote hydraulic ram (Fig 111).

On wide drills, two lift units are fitted, each one operating half of the coulters. This avoids wasteful double drilling when only a narrow strip of land remains to be drilled. These so-called self-lift coulter mechanisms aim at eliminating the need for a second man, but there are times when a second man can justify his labour cost by speeding up the filling of the drill and by preventing coulter tube blockages continuing unnoticed, especially when combine drilling and when the fertiliser is in poor condition.

The coulter lift mechanism is linked with the feed mechanism drive so that the latter is engaged automatically as the coulters are lowered.

Construction and working principles of combine drills

Reference has been made in Chapter 3 to the use of artificial fertilisers in crop production. The combined grain and fertiliser drill was introduced in an endeavour to obtain the maximum benefit from fertilisers applied to cereal crops. When fertiliser is applied to the seedbed before drilling, some of the valuable elements are invariably lost by evaporation and *leaching* before the growing crop can utilise them. So it is advantageous to place these supplementary plant foods into the drill together with the seed. There are, however, important qualifications, which are mentioned later. So effective is this technique that the application rates of the fertilisers can—indeed must—be reduced, and this can produce a considerable saving. Also, the more rapid establishment of the crop can have marked effects on its yield. Some claim that time and fuel is saved by the distribution of both seed and fertiliser in one operation, but on the other hand, extra time is needed to load the drill with fertiliser when drilling may be a matter of urgency.

Disadvantages of the combine drill are its higher initial cost, its increased weight and draught by comparison with the standard grain drill (an important consideration on heavy wet land), and its rate of deterioration by the corrosion it invariably suffers.

A risk in 'down the spout' fertiliser application is the danger of impairing the germination and depressing the development of the crop by the application of excessive quantities of fertiliser with the seed—particularly with fertilisers of the nitrogenous and phosphatic types.

The construction of the combine drill is similar to that of the grain drill already described. However, its hopper is divided into two sections, one for seed and the other for fertiliser (Fig 112). The fertiliser section of the hopper also has its own fertiliser-feed mechanism, usually of the rotating *star-wheel* type (Fig 113). The star wheels are driven from the axle of the drill through a chain and/or gear drive. The fertiliser application rate is usually controlled by an aperture slide which restricts the flow of the material from the hopper to the coulter tubes.

The star-wheel type of feed mechanism cannot handle even the smallest lumps in the fertiliser, so only free-flowing material

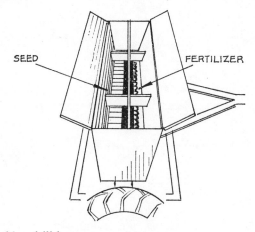

112 Combine drill hopper

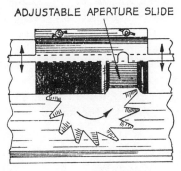

113 Dispenser unit commonly used in the fertiliser section of the combine drill hopper

139

should be used if even application is expected. On some drills, the fertiliser is delivered to the coulters through the same tubes as the seed. In this case, any blockage of coulter tubes by lumpy fertiliser also prevents the seed reaching the drills. This often accounts for the all too conspicuous gaps seen in drilling when the plants emerge. To prevent this, some drills have two sets of coulter tubes, one for the seed, and the other for the fertiliser delivery (Fig 114).

114 Independent grain and fertiliser tubes

ATTACHMENT TO TRACTOR

When hitched to the tractor, the frame of the drill should be parallel to the ground, or the coulter depth adjustment may be restricted. Also, as the lengths of the coulter frames vary, the working depths of the coulters may vary as well. A trailed drill is levelled by selecting the appropriate vertical position for the hitch clevice. A strong draw pin should be used as the weight of the drill with its load of seed and fertiliser, and possibly an operator on its platform, can be considerable.

When attaching the mounted drill, the level of the frame is controlled by the length of the top link of the three-point linkage.

Operational adjustments

Reference has been made to the more usual operating controls and adjustments provided on drills, but these need to be considered in more detail, for although the adjustments are few, they are most important.

SOWING RATE

Various methods of adjusting the sowing rate on different drills have been mentioned. The method common to most drills is the variation of the speed of the rotating element in the feed mechanism. Sliding fluted rollers on external force-feed drills and the use of alternative sides of the dual-faced internal force-feed rotor are other ways of varying the rate. The settings recommended by the manufacturer can be relied upon for reasonable accuracy, although the seed of different strains of any one cereal can vary considerably in size, and to some extent in shape. So, for absolute accuracy with an unfamiliar seed, a calibration test should be carried out to check that the setting is correct for the particular seed to be sown. This may be considered a time-wasting performance by many tractor drivers, but one can often see evidence in fields of drastic sowing rate alterations which have resulted from 'trial and error'!

STATIONARY CALIBRATION PROCEDURE

Some manufacturers supply detailed instructions for stationary calibration of their drills. Where instructions are not available, the following procedure may be adopted:

1. Put a small quantity of the seed to be sown into the hopper, and spread it out so that all the apertures to the feed units are covered.
2. Set the sowing-rate lever on its graduated scale to the position indicated for the proposed sowing rate. If no indication is given on the drill and the instruction book is not available, select a random setting and then modify if necessary in the light of the check herewith outlined.
3. Place a canvas sheet or other means of collecting grain from the coulters beneath the coulters and lower them on to it. With a self-lift drill, the weight of the coulters must be relieved with a lever or spanner before the trip is released, or the impact of the coulters

striking the ground may cut through the sheet and even damage the coulters themselves if the floor is of stone or concrete.

4. Jack up one wheel of the drill, and turn it through the number of revolutions it would make in drilling a specific area. (One-tenth of an acre is a convenient area on which to base the check.) The following formula gives the number of revolutions the wheel would make in covering one-tenth of an acre:

$$\frac{\text{Number of sq ft in one-tenth acre}}{\text{Drill sowing width (in feet)} \times \text{Wheel circumference (in feet)}}$$

If the drill sowing width is 10 ft and the wheel circumference is 12 ft, the wheel revolutions necessary to cover one-tenth of an acre (or 4,356 sq ft) will be:

$$\frac{4,356}{10 \times 12} = \frac{363}{10} = 36 \cdot 3$$

Turning the drill wheel $36\frac{1}{3}$ times represents drilling a tenth of an acre.

5. Turn the drill wheel this number of times and take up and weigh the grain collected on the sheet. Multiplying this weight by 10 gives the weight of seed the drill would sow per acre at that particular setting. If the amount is greater or less than the sowing rate required, the setting on the feed mechanism must be modified and a further check made, and so on until the correct setting is ascertained. Once a drill has been calibrated with a particular variety of seed, the setting should be recorded for future reference.

As the sowing rate of cup-feed drills is affected by vibration, the stationary test is not so accurate for that type of drill, although it may be used as a guide.

FERTILISER APPLICATION RATE

The fertiliser sowing rate on combine drills is usually adjusted by a lever which controls a series of restrictor slides at the fertiliser outlet passages (Fig 113). Calibration checks are rarely made on the fertiliser distribution side of the drill, as the condition of the fertiliser or even the humidity of the atmosphere can influence the

5. Toolbar with coiled spring tines

6. Three-row potato planter with pallets

7. Crop sprayer with band-spraying drop tubes

8. Effect of band-spraying on row crop in dirty land

rate of distribution considerably. Because of this, the drill manufacturer's indicator scale at the control lever is usually accepted.
In any event, inaccuracy of the fertiliser discharge mechanism is
not so serious as that of the seed delivery mechanism. However, it
is important to keep a close watch on the drill acreage meter, and
on the fertiliser used, to avoid serious discrepancies.

SOWING DEPTH

Depth of sowing is another important setting which seldom
receives the precision it deserves. Its object is to adapt the drill to
the requirements of the seed and the condition of the seedbed.
Too often no account is taken of the firmness or hollowness of the
seedbed, or the earliness of lateness of the operation, etc. The
work should be closely examined at intervals to check the depth of
sowing. Remember that the headland soil is usually firmer than
that in mid-field, and so the inspection and any necessary correction should be repeated when the headland drilling has been
completed.

SETTING MARKERS

Drills of up to 15 coulters are sometimes used without markers.
Instead, the tractor driver matches his bouts by estimating the
distance between the end coulter and the previous bout. Although
some skilled men can produce acceptable work by this method,
many fall short. Much greater accuracy can be achieved, and with
less concentration, by the use of markers. These consist of telescopic or otherwise extensible arms, attached to the sides of the
drill and projecting outwards. Each arm carries a tine or disc
which draws a fine marking groove on which the operator positions
his appropriate front tractor wheel when drilling the subsequent
bout. These markers require careful setting if 'missing' or 'overlapping' is to be avoided. A convenient procedure is as follows:
1. If the drill is of the trailed type, it should be attached centrally
on the tractor drawbar.
2. If the outfit is now driven 3 or 4 yd along the headland of the
prepared field with the tractor steering straight, the tyre im-

pressions made on the soil can be used for the following measurements.

3. The distance should be measured between the impressions made by each front wheel of the tractor and the outermost coulter of the drill on the same side. In fact, it should only be necessary to carry this out on one side of the outfit, as the other side should be identical.

4. To this measurement should be added one coulter width—7 in on most drills—to give the precise distance the marking disc or tine must be from the end coulter of the drill (Fig 115). If the

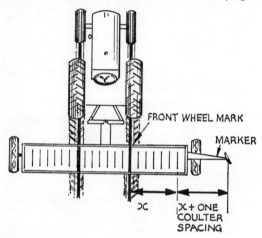

115 Setting drill markers

additional coulter width is omitted, double drilling will occur on the end drill of each bout. The two marker arms are interconnected with a rope or chain, and arranged so that the tractor driver can bring the nearside and offside markers into use alternately at the commencement of each bout.

ACREAGE RECORDER

The acreage recorder is a small unit which comes into operation as the coulters are lowered and the seed-sowing mechanism engaged. It records with reasonable accuracy the area of land drilled. Needless to say, the meter should be set to zero before starting work.

WHEEL SCRAPERS

Wheel scrapers are fitted on steel wheeled drills to remove soil from the rim immediately there is any tendency for it to stick. They should be set as close as possible to the rims without touching, or their effect will be worthless, as one layer of soil on the rim will continually pick up further soil—sometimes with seed sown on the previous bout.

Field procedure

The direction of the drilling, and the arrangements for the supply of seed, and of fertiliser if necessary, must be determined before work begins. In view of the urgency of the work, it is often justifiable to have a second tractor standing by with a trailer carrying the necessary supplies.

Many farmers prefer to drill their fields in bouts. In this case, a wide headland is first drilled, the outfit making five or six circuits starting at the field boundary and continuing until an ample margin has been drilled around the field. This should be wide enough for the outfit to come out of one bout and enter the next straight, and without too small a turning radius. Adequate width is necessary for accurate matching of the bouts at the ends, and to avoid undue disturbance of the drilled headlands. The inside wheel-mark of the drill produced on the last round of the headland drilling serves as the guide line for consistent lowering and raising of coulters. When the headland drilling has been completed, it is usual on a level field for the drill to travel in line with the longest hedge or fence, working its way across the field by adjacent bouts.

Straight drilling is desirable, though not imperative for cereal crops. Accuracy of matching up one bout with the next is important.

In some parts of the country, the round-and-round method of drilling is more commonly used, i.e. the outfit follows the field boundary and continues to work until the centre of the field is reached. Although this involves 'loop' turning of the outfit at the corners after the first few circuits have been made, it is still a very quick and efficient method of drilling, and one applicable to small as well as to larger fields. With this system of drilling, a second tractor with its loaded trailer is usually drawn progressively across a diagonal of the field for reloading.

Unnecessary traversing of the unsown seedbed prior to drilling should always be avoided, particularly if the soil is moist. Even the compaction caused by the tractor wheels during the normal drilling procedure may have an adverse effect on the drills which follow the tracks.

It is good practice to trail a set of seed harrows behind the drill to ensure complete coverage of the seed, and so eliminate a separate and often uneconomic operation. Since the introduction of safety regulations for farm machinery, however, it is illegal for a man to stand on a platform in front of a trailed implement. This is avoided by the use of a very light spring-tine harrowing attachment that can be fitted beneath the platform as an optional extra. When the drilling is performed single-handed by the tractor driver, the ruling does not apply.

Having outlined the general procedure of work in the field, a few practical tips may be added for the less experienced operator.

As the drill enters each bout, the coulters should be lowered at least a yard before the headland mark is reached. This allows time for the seed to reach the coulters from the feed units. For the same reason, it is advisable to broadcast a handful of seed in front of the drill coulters whenever an emergency stop has to be made during work.

A constant watch should be kept on the coulter tubes for blockage. The point has been made earlier of the value of a second man riding the drill to keep his eye on the seed and fertiliser runs.

When working on hillside land, a watch also needs to be kept on the seed in the hopper, as this may tend to shake down to the lower end of the hopper—despite the dividing panels usually incorporated—allowing some of the seed outlets to run short.

Another tendency on some combine drills is for the fertiliser to work to one end of the hopper, even when drilling level land. This is due to the action of the star wheels which all rotate in the same direction, with only half of their surfaces exposed to the material, resulting in a slight cross-conveying action.

Reference has been made to the importance of depth control when drilling, and this is another point which the second man can keep under observation as work progresses. When changing from one type or strain of seed to another, the drill hopper should be thoroughly cleared.

Maintenance

DAILY ATTENTION

The land-wheel axle bearings, the sowing-mechanism spindles, the disc coulters and the self-lift mechanism, where fitted, should normally be lubricated every day.

At the end of the day's work, the drill should be completely covered, and any soil adhering to the coulters removed— particularly if overnight frost is likely. All the fertiliser in the hopper of a combine drill should be used up before stopping for the night. Even then, it is advisable to turn the fertiliser mechanism over by hand before work restarts the next morning. A special handle or spanner is sometimes provided for this purpose. If the combine drill is likely to be out of service for more than a day or two, the fertiliser section of the machine must be thoroughly cleaned out to prevent corrosion.

PERIODIC ATTENTION

Before work is started at the beginning of the season, the general condition of a drill should be checked, and if the machine was stored away correctly it will be necessary to remove rust preventive from the working parts. The fertiliser distribution section will need to be reassembled and checked for free operation.

The tensions on the coulter springs should be checked for equality by lifting each coulter frame by hand in turn, and assessing the resistance of its spring. On most makes of drill a simple adjustment is provided to equalise the tensions so that all coulters sow at the same depth.

Another check, which should be made on drills having done a few seasons' work, is on the spacing of the coulters. If the coulter frames are loose or have end float on their forward pivots, the spacing will be uneven. Sometimes the frames become distorted as a result of the drill being reversed in the field with its coulters lowered.

END OF SEASON ATTENTION

At the end of a season's work, the drill hoppers should be cleaned out, and the fertiliser feed mechanism components removed for cleaning and anti-rust treatment. To facilitate this, the fertiliser-dispensing mechanism is usually readily accessible and easy to dismantle. The coulter tubes should also be removed and cleaned, and if of spring steel, they too should be protected from corrosion. Increasing use is being made in combine drills of corrosion-resistant materials. Where this is the case it is as well to follow the maker's guidance as to cleaning procedure. The time required to clean and protect the coulters is amply rewarded by the absence of soil adhesion at the start of the next season's work.

Chains and gears should be closely examined and adjusted if necessary. Sprockets should be checked for excessive wear. On steel-wheeled drills, adjustment is usually possible at the land-wheel bearings to correct end float. Pneumatic-tyred wheels seldom require bearing adjustment, but the tyres need inspection and the care advised for all tyres, i.e. correct operating pressures, relief from implement weight during storage, and protection from continuous exposure.

Safety precautions

1. Always use a close-fitting draw pin with retaining device.
2. Ensure all guards are in place.

3. Where a drill man is carried, special regulations about handrail and footboard should be followed.

4. Where a seed agitator is fitted, avoid catching fingers when clearing the hopper.

5. Avoid abrupt starting and stopping when an operator is carried on the drill.

Construction and operation of unit-type spacing drills

Although not so extensively used as corn drills, spacing drills designed for sowing such seeds as sugar beet, fodder beet, mangold, swede, kale and maize, etc., have been the subject of considerable research in recent years. Many corn drills of the cup feed and external force-feed types can handle root and kale seed with reasonable accuracy, but better results are obtained with drills having a low hopper position, to give a regular seed discharge, and a technique known as *precision drilling*. A choice of dispenser elements in the feed unit enables the drill to be adapted for sowing seeds of widely varying sizes and at different spacings.

Precision drilling has the advantages of considerably reducing the required sowing rates, eliminating competition in the early stages of plant growth, and simplifying the down-the-row thinning technique of plant population control (see Plate 4).

Unlike grain drills, unit-type spacing drills do not share a common hopper. Instead, each metering unit has its own hopper together with its own provision for opening up the soil and covering the seed after its discharge (Fig 116). The drive to the feed mechanism on the seeder illustrated is provided by the front land wheel. Some seeders are p.t.o.-driven to give more positive drive and to ensure regular seed delivery when seedbed conditions are not favourable. Provision is made for automatic cut-out of the power drive when the drill is raised at the bout ends.

Some makers offer an electric warning lamp circuit to indicate when any of the units become inoperative. Some seeder units can be attached to the standard three-point linkage tool bar, while others are designed for a specific tool bar which has provision for

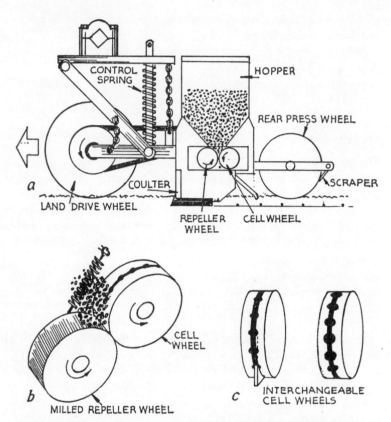

116 Precision seeder unit: *a* General construction, *b* Cell wheel and repeller wheel, *c* Alternative cell wheels

quick and easy change of row-width setting (Fig 117). In an endeavour to improve the precision of these drills, many different types of feed mechanism have been introduced. Two types are shown in Figs 116 and 118, one employing a *cell wheel*, and the other a horizontal perforated belt. Various methods of ejecting the individual seeds from the cell wheel have also been devised.

One difficulty encountered in single seed dispensing is the lack of uniformity of size and shape of some kinds of seeds. Two answers to this problem are *rubbing* and *pelleting*.

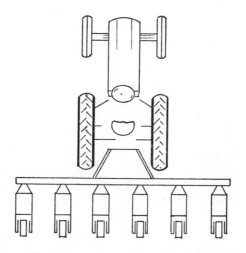

117 Unit-type seed drills

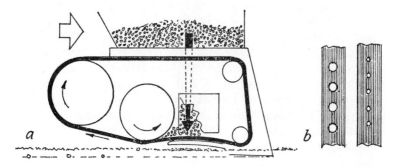

118 Perforated-belt feed mechanism: *a* Feed system, *b* Sections of alternative belts

Both these treatments have been used with sugar-beet seed. In the rubbing process, which is quite widely practised, the corky 'fruits' are subjected to a delicate grinding action until they are reduced to a specific size. A commonly used grading is that which passes through an $\frac{11}{64}$-in screen but is retained on a $\frac{7}{64}$-in screen. Besides giving greater uniformity of size and shape, this process

151

also reduces the number of potential plants produced by each individual fruit, and gives a higher proportion of mono-germ seeds, i.e. seeds which produce one plant only and thereby reduce the work of thinning. Seed treated in this way is usually referred to as *rubbed and graded*, whilst the normal untreated seed is referred to as *natural*. Sometimes, root seed is graded by gravity, as uniformity of weight is another factor which can influence the regularity of seed spacing from spacing drills.

The alternative, though less widely adopted, process of pelleting involves coating the seed with a substance to achieve a uniform size. The coating must be tough enough to resist shattering during transport and handling, and yet soluble enough to dissolve readily in moist soil—two requirements not easy to reconcile. Where too soft a casing has been used, the inaccuracies in the metering of the seed have been greater than with natural seed. Where the casing has been excessively hard, germination of the seed has been seriously delayed, particularly where the drilling has been carried out on a fairly dry seedbed.

The individual unit principle of construction gives a wide range of adjustment of lateral drill spacing. Unless a special narrow-tyred row-crop tractor is available, the row width will be determined by the need to accommodate the larger tyres of the general-purpose tractor. So far as the sugar-beet crop is concerned, a 20 in row width is commonly used, and has proved to be satisfactory for drilling, for after cultivations, and for harvesting operations. Methods of varying sowing rates, and of interchanging alternative cell wheels on different machines, are too diverse and specific to generalise, but the manufacturer's guidance is normally adequate.

Ground speed is an important factor in the efficiency of precision seeder units. Generally, it has to be kept down to about 3 miles per hour, although it does depend upon the condition of the seedbed and the required precision of seed placement. The seeder units, although positively attached to a tool bar or drill frame, are free to float and follow ground contours (Fig 116). Adjustment is sometimes possible on the coulters which open the narrow furrows to receive the seed. After the seed has been delivered, the furrow is closed by deflectors or a drag-chain, which replace the soil. The seed is often discharged from the unit only an inch or so above the furrow, and this gives more accurate seed spacing by preventing excessive rolling. The rear rollers fitted to seeder units perform the final operation of lightly compacting the soil around the seed. It is sometimes possible for these rear wheels to be hinged upwards out of their working position to avoid picking up moist soil or when their use is likely to cause capping. When these roller wheels are in use, their scrapers should be set close to their rims.

Accurate setting of markers is vital, to facilitate any later inter-row hoeing, down-the-row thinning and the eventual topping and lifting operations. A stabiliser fin fitted to the tool bar or drill frame helps the units to run true and provides a guiding groove to align the tool bar for hoeing. Some form of control spring is usually provided on each unit to prevent bouncing, and this needs to be adjusted to suit the working conditions. Where such crops as sugar beet and carrots are to be harvested mechanically, an adequate width of headland should be drilled.

A *band-sprayer* attachment may be fitted to the drill assembly for simultaneous spraying and drilling. This is a technique of chemical weed control known as *pre-emergence*, and it is described more fully in Chapter 5. One nozzle is situated behind each seeder unit, and adjusted to spray the chemical on the band of soil strad-dling the row (*drill*) of seeds. As a result, the soil immediately adjacent to the seedling plants when they emerge is completely free of weeds, and will remain so for several weeks.

Construction and operation of grass-seed drills

Although grass seed and grass and clover mixtures can be sown with external force-feed drills, the work is usually done better with a drill designed exclusively for them, particularly where the sowing rate is very low or very critical.

The drill is usually light, and may be designed for attachment direct to the tractor three-point linkage, or as an attachment for a tractor tool bar, or as an attachment for a roller. Some manu-facturers supply grass-seed attachments for their corn drills. When used with a ribbed roller, the chinks made in the soil by the

rings are often sufficient to introduce the seed into the soil. The
feed mechanism may consist of serrated rollers, serrated wheels or
revolving brushes as shown in Fig 119.

The drive for the mechanism is often in the form of a floating
cleated wheel driven by ground contact. Many of these drills have
no coulters, but instead scatter the seed loosely on the surface.
For *undersowing*, i.e. sowing clover/grass-seed mixture on soil
already supporting a growing cereal crop (so that the herbage has

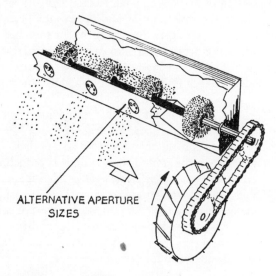

ALTERNATIVE APERTURE
SIZES

119 Brush-feed grass seeder

become established by the time the cereal crop is harvested), the
lack of coulters is no disadvantage. For the direct seeding of uncropped land, however, coulters can be an advantage. In some
cases, the technique of double drilling is used, whereby the drill
sows half the quantity of seed in one direction and the other half
at right angles. This practice facilitates the separation of the seeds
when mixtures have to be sown, so that the optimum depth for
each can be used.

SWARD INCISION SEEDER

Fig 120 illustrates another unique type of drill which is fitted with robust spring-loaded tines and sharp knife coulters. These cut slits in established grassland to introduce new seed into the soil as a means of improving the herbage. The slitting of the turf is itself

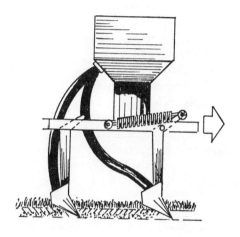

120 Grass seeder. Note the robust spring-loaded tines and sharp coulters

beneficial to the sward, whilst good account has been given of the combined operation on poor pastures so long as the existing turf has been controlled, pending establishment of the new herbage. The machine has also successfully established ryegrass on cereal stubble.

Construction, operation and maintenance of potato planters

The mechanisation of the potato crop has presented many problems. Some of these have been more or less overcome, but the mechanisation of the planting and harvesting of the crop has not reached the same level of standardisation as that of other crops. Some of the problems spring from the fact that potatoes vary both in shape and size, and that they are very susceptible to mechanical damage. So far as potato planting is concerned, avoidance of damage to the tuber is of particular importance where *chitted* seed is used. Chitting, now quite widely practised, is the storage of the tubers in controlled conditions conducive to the development of strong and even shoots in readiness for planting, the advantages being quicker establishment and earlier maturity of the crop. When using such seed, it is important to avoid any harsh handling which may damage the shoots.

Potato planting is tackled in various ways on different farms. Where very small acreages are involved the actual planting is done by hand, and the *ridging* (opening up of the trenches) and *splitting* (covering of the tubers) are performed with a tractor-drawn ridger. Where more intensive growing is practised, the planting operation may be carried out with varying degrees of mechanisation. There are many planters on the market, and although these vary in pattern and in the amount of manual participation they require for maximum output, they are broadly classified as either *semi-automatic* or *automatic* types. In either case, ridging bodies or concave discs may be used to open up the trenches and cover the tubers. The classification refers to the actual tuber-dispensing mechanism.

SEMI-AUTOMATIC PLANTERS

Fig 121 illustrates the layout of a machine of this type designed as a three-row planter, and Fig 122 shows the manual loading of the dispenser wheel. Another type employs segmented magazines which the operators keep charged—one tuber to each compartment. The magazine rotates and as each segment lines up with the mouth of a vertical planting tube, the tuber falls down to the open trench (Fig 123). Where chitted seed is used it is preferable for the tubers to be taken straight from the trays or pallets on which they are transported from their storage place. For this reason some makers provide an attachment to accommodate the pallets as an alternative to the standard hopper (Plate 6).

121 Semi-automatic three-row planter

122 Operator charging tuber dispenser wheel

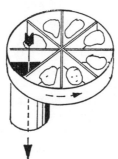

123 Rotating magazine type tuber dispenser

AUTOMATIC PLANTERS

With most automatic planters it is advisable to have someone constantly supervising the dispensing mechanism and correcting any occasional *misses* in the delivery mechanism, particularly if the seed has not been accurately graded. One three-row automatic planter is illustrated in Fig 124 and its principles are self-explanatory. Fig 125a illustrates a machine which incorporates a corrector mechanism. If one of the trays on the vertical conveyor fails to collect its tuber, a sensor finger drops and brings into operation a drive to the auxiliary magazine which rotates and discharges a tuber into the planting tube. The numbers on the illustration (Fig 125b) indicate the chain of events.

OPERATION

The operating controls on potato planters—apart from the usual hydraulic lift—are usually simple levers provided to engage the drive to the feed mechanism, and, on the automatic planters, to release the tubers. Operational adjustments, on the other hand, vary considerably, although they have the same basic functions. These follow with common adjustment ranges:

1. Depth of planting (4 to 7 in).
2. Width of row spacing (24 to 36 in).
3. Tuber spacing within the row (9 to 18 in).
4. Width of ridger mouldboards, or angle of discs (whichever are used).

The optimum setting for each of these factors depends on the nature of the soil, the variety of seed, the time of planting, etc. Most machines can be set to plant land which has been previously ridged, or land directly following final cultivating operations. In the latter case the fitting of opening ridger bodies is, of course, necessary. Markers must be set accurately when planting unridged land, and the procedure has been outlined in connection with ridgers in Chapter 2. Some machines can accommodate a fertiliser attachment, so that this important operation can be carried out at the same time as the tubers are planted. Some machines place the fertiliser on either side of each row of tubers, so making the plant nutrients readily available without the danger of direct contact. Before investment is made in such an attachment, however, consideration must be given to the weight factor. A three-row planter carrying seed and operators may be a considerable

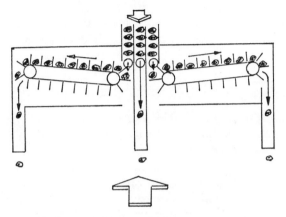

124 Automatic three-row potato planter

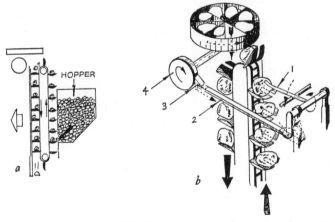

125 Automatic planter. *a* With elevator trays to single out and discharge tubers. *b* Automatic corrector mechanism: when a tray is not loaded, sensor finger (1) drops, pawl stop (2) is withdrawn and pawl (3) rotates with driving flange (4) to rotate magazine one sector and drop the required tuber

load on the tractor, especially on hilly land, and in such circumstances the additional weight of the filled fertiliser attachment might prove to be more of a drawback than an asset.

For efficient working of automatic planters, good grading of the seed is important, particularly with the long varieties of potatoes. Ideally, rounded seed of consistent size should be used. When chitted seed is to be planted, the shoots should preferably be short and green.

MAINTENANCE

As the selection and discharge systems on current planters differ considerably, it is not possible to generalise about lubrication. Operators should acquaint themselves with all moving parts of their machine, and follow instructions given by the manufacturers.

Maintenance adjustments also vary. As a seasonal machine, the potato planter can suffer seriously from corrosion, especially if fitted with a fertiliser attachment. So, after a season's use, it must be cleaned and rust-prevention measures taken.

5 FIELD CROP-SPRAYING MACHINES

For centuries, it has been known that some substances are toxic to growing vegetation, and in the latter part of the last century it was discovered that some chemicals were selective in their action on plants, or, more accurately, that the susceptibility of certain plants to particular chemicals varied considerably. This phenomenon aroused some interest in the possibilities of chemical weed control in agriculture and horticulture, but comparatively little progress was made until the 1940s, when increased productivity was a matter of extreme national urgency.

Since that time, the crop sprayer has become more and more commonplace on the farm, and today it is recognised as an essential item of equipment. Its arrival opened up new horizons for the large scale use of insecticides and fungicides which, although in existence for many years, were mainly limited in their applications to vegetable, flower and fruit growing.

Applications of the crop sprayer

CONTROL OF WEEDS

The importance of good ploughing and cultivating techniques to destroy or suppress weeds has been stressed in earlier chapters. However, there are times when even the best management of arable land fails to eliminate entirely weed competition. The presence of certain weeds may be due to unsatisfactory soil conditions which cannot be quickly remedied. Mechanical control measures in growing crops are, of course, restricted to row crops, except for the rather limited practice of surface scarifying of field crops with harrows or weeders. So it is in close-growing crops that selective weed killers are particularly advantageous. The extensive range of spray materials now available, and advances in application techniques, have considerably widened the scope of culture by chemicals. Significant progress has also been made in the degree of selectivity attainable.

The sprayer simplifies the control of weeds in growing cereals, and so achieves higher yields and eases the harvesting operations; indeed, it may be said to complement the combine harvester. It also indirectly controls insect pests by eliminating those weeds which act as host plants to insects. Certain fungus diseases harboured by weeds can also be restricted. The control of weeds in grassland is a further valuable application of herbicides.

In spite of the important achievements attained in the field of chemical weed control, spraying should really be regarded as complementary to good management and efficient tillage, and not as a substitute for either.

CONTROL OF INSECT PESTS

Insecticides (or pesticides, as they are sometimes called) have developed side by side with herbicides, and are now widely used for the destruction of parasitic insects which attack growing crops directly or indirectly. The application of the appropriate chemical may have to be directed at either the crop foliage or the soil.

CONTROL OF FUNGUS DISEASES

The use of the sprayer for the control of fungus diseases is rather more limited in its application, although one common operation is spraying for the control of potato blight.

DESTRUCTION OF POTATO HAULM

The destruction of potato haulm simplifies harvesting operations, and is particularly important if the haulms of a potato crop have become contaminated with the spores of potato blight when the only satisfactory treatment is the complete elimination of the diseased foliage to avoid infection of the tubers. Special chemicals are available for this purpose.

SUPPLEMENTATION OF PLANT NUTRIENTS

Sprayers may be used to apply one or more of the principal plant nutrients as a liquid, but they are more usually employed for the application of trace elements. As the word *trace* implies, these elements need to be present in only very small quantities, but they are nevertheless vital for healthy plant growth. Consequently, for this type of work the sprayer should be used only under the guidance of a soil expert, with facilities at his disposal for accurate soil analysis. Serious chemical imbalance of soil can result if such work is done without technical guidance. Applications based solely on the appearance of disease symptoms in a crop can entail risk of crop failure and soil damage.

Alternative spraying techniques

The efficiency of a sprayer in these various applications may be influenced by several factors, such as stage of crop development, density of crop, nature and population of weeds, pest infestation, soil type and condition, and weather conditions. These variables, together with the alternative types of chemicals available, demand different rates, times and forms of application.

TIMES AND TECHNIQUES OF APPLICATION

For best effect, it is vital that the weeds, pests or fungi are attacked when they are most susceptible, and if this coincides with a time when the land is uncropped, or when the crop is least susceptible to physical and chemical disturbance by the treatment, so much the better. Timing of the spraying operation is distinguished by the following terms: *pre-sowing*, *pre-emergence* and *post-emergence*.

Pre-sowing techniques

Although not so widely practised as other techniques, the pre-sowing method of weed or pest control sometimes has considerable advantages. The diluted chemical may be applied to stubble or to other uncropped land, either before ploughing or between cultivating and sowing operations. Good results have been obtained in the control of weeds and certain soil pests in this way. Trace-element deficiencies may also be made good at such times.

Pre-emergence technique

In the pre-emergence method, the herbicide is applied to the soil, either simultaneously with the drilling operation, or immediately after the drilling so that the seedlings emerge in a weed-free environment. This technique is commonly used for sugar beet, as the seed is relatively slow in germinating, and without such control measures, a considerable population of weeds may become established by the time the seedling plants emerge. Small spraying attachments are available for fitting to drills, toolbars and seeder units. The nozzles are positioned behind each drill coulter or seeder unit to spray a band about 7 in wide straddling the seed drill. The liquid is discharged at low pressure, although the pump operates at sufficient pressure to agitate the contents of the tank. Some attachments for band spraying have the tank at the front of the tractor to avoid obscuring the operator's view of the nozzles. Band spraying is still the subject of experiment. Its effectiveness is influenced by the chemical and physical conditions of the soil, and a soil analysis is a prerequisite of the treatment. Weather conditions also have a bearing on its reaction (Plates 7 and 8).

Post-emergence technique

The majority of spraying is carried out by the post-emergence technique, i.e. after the crop has emerged. The treatment, however, should not be delayed too long, or the weeds—if these are the target—may have passed their most susceptible stage of growth. It is even more difficult to time the treatment for the control of insect pests. The technicalities of spraying growing crops are discussed later in the chapter.

Grassland spraying

Although good grassland management should generally avoid the need for weed control, the sprayer can make a valuable contribution to the improvement of a neglected sward. Here, again, the success of the operation depends largely upon recognition of prevailing soil conditions, identification of weeds and correct timing of control measures. Although most of the chemicals used for grassland spraying are not poisonous to animals, some tend to make toxic weeds which they would not normally eat palatable to stock. So, there is a risk of poisoning or milk tainting if sprayed pasture is grazed too soon after the treatment.

A more recent application of the sprayer on grassland is the destruction of the old herbage in conjunction with the introduction of new seed by means of a sward seeder or similar machine.

Action of spray materials

Chemical analysis of spray substances is beyond the scope of this book, but operators of spraying machines should have a little basic knowledge of some of the effects the more common spray materials have on plants and pests.

HERBICIDES

Herbicides are available in various forms, each having certain limitations, and so needing careful selection for particular treat-

165

ments and techniques. Two groups commonly used are the *contact* type and the *translocated* type.

Contact herbicides

Contact herbicides destroy or suppress weeds by scorching their foliage and growing points. The selectivity of their action depends mainly on the physical characteristics of the various plants on to which they are sprayed. For instance, weeds having broad leaves and exposed growing points are generally more susceptible to toxic chemicals. Cereals, on the other hand, have concealed growing points, and the liquid sheds quickly from their slender foliage, so that they are able to resist the toxic effect at normal application rates. As a result, when a field of growing corn is sprayed, certain weeds are destroyed, while the crop remains unharmed, or at worst only very slightly affected by the treatment.

Translocated (or growth regulating) herbicides

Translocated-type substances act in a different way. The chemical, on making contact with the foliage of certain weeds, is rapidly absorbed and translocated to all parts of the plant's system, including the roots. The effect on a susceptible weed is a rapid but brief period of contorted growth followed by cell destruction. The selectivity of the chemical action results from differences in cell reaction to the poison. It is this translocating process which makes *growth-regulating* type weed killers (as they are sometimes called) so effective in the control of many deep-rooted perennial weeds, without the need for complete saturation by high-volume spraying. Most preparations in this group can be applied at medium or low volume rates. Other noteworthy advantages are that many of the chemicals in this group are non-toxic to the human system and to stock, and that they are available as emulsions which are easy to mix with water.

INSECTICIDES

Good husbandry, by way of sound rotations and careful seed-variety selection, can reduce pest damage as well as weeds. Even so, there are times when some form of chemical treatment may be desirable, if not imperative.

Intensive research by entomologists into the life cycles and habits of insect pests has suggested several approaches to the problem. These include the application of chemical dusts to the soil, the dressing of seed, and the application of insecticides with the crop sprayer. Insecticides have been classified under the headings of *contact* type, *stomach-poisoning* type and *systemic* type. The first two usually overlap in their effect, and it is convenient to consider them together.

Contact and stomach-poisoning insecticides

These two types of insecticides destroy the insects by direct contact, by internal poisoning—as they eat the foliage coated with the chemical, or by both these forms of contamination.

Systemic insecticides

Some common pests are not seriously affected by contact-type chemicals, but most of these are susceptible to the systemic type. In this case, the chemical is absorbed by plants through their foliage and flows with the sap through the plants' systems. Insects which survive the actual contact of the chemical are subsequently poisoned as they suck the plant juices. Application rates vary widely, and instructions are always supplied with the chemical preparation. These should be strictly followed, as some poisons have a strong residual effect.

FUNGICIDES

The most common use of fungicides is protection of the potato crop from blight. High-volume application is used, and drop legs are fitted to the boom to spray the underside of the foliage in order to cover it completely. Even after this treatment, it is still advisable, before lifting the crop, to destroy the affected haulms completely with a powerful chemical.

PLANT NUTRIENT SPRAYS

Although for the sake of completeness this is included under the general heading of 'Action of spray materials', the application of such substances with the spraying machine is so specialised that nothing can usefully be added to the reference made earlier.

So far, in this chapter, spray materials have been classified by their functions. Many chemical compounds are in use, and they are continually being superseded by others, with almost bewildering rapidity. For this reason, it is felt that little useful purpose would be served in attempting any chemical classification, especially as the producers of spray preparations are always willing to supply such information.

Physical classification of spray materials

An important form of classification for the operator of the spraying machine concerns the physical properties of the various preparations as these have a bearing on mixing procedure, agitation requirements and indeed on the mechanical specifications of the machines required to apply them.

The concentrated chemicals used for spraying are marketed in various forms and strengths. Because of this, their mixing and dilution requirements differ. Also, they require varying degrees of agitation to keep the liquid adequately mixed in the tank. Most manufacturers issue detailed instructions regarding mixing and agitation, but the following classification may be of interest.

SOLUTIONS

Whether the concentrates are supplied in liquid or powder form, solution type chemicals readily mix with water and form homogeneous liquids which require little or no pre-mixing or agitation.

SUSPENSIONS

Suspensions may be supplied as a paste or wettable powder.

Chemicals of this type are in the form of finely divided solid particles which are insoluble in water. So to ensure consistent chemical strength, the liquid must be agitated continuously to prevent the solid particles settling at the bottom of the sprayer tank. For this reason, suspension-type sprays should be used only in sprayers having mechanical agitators. These usually consist of revolving paddles (Fig 126).

EMULSIONS

Many emulsions are mineral oil derivatives and usually have to be pre-mixed, that is diluted initially with a small quantity of water before mixing with the bulk of the water.

CORROSIVE FLUIDS

Corrosive chemicals should be used only in spraying machines specially designed for them; that is, drip proof and possessing components with a high resistance to corrosion. Operators must take special care to reduce the risk of personal injury, from physical contact with the liquid or by inhaling the vapour. The special safety measures required are included in *Safety precautions* at the end of the chapter.

VOLATILE OILS

Although volatile oils have limited application on farm crops, they are occasionally used to good effect. However, the damaging influence of oil on rubber hoses, and the risk of fire from the atomised oil being discharged from the nozzles, must be borne in mind.

WETTING AGENTS

The waxy surfaced foliage of certain weeds causes spray liquid to run off freely in beads, reducing its effect. This may be overcome by incorporating a wetter in the liquid, which has the effect of

169

reducing the surface tension of the liquid and so causing it to spread on the foliage and give more complete coverage.

Types of crop sprayer

Spraying machines may be classified by: (a) system of liquid delivery—hydraulic or pneumatic, (b) volumetric capacity— high volume, medium volume, low volume, or universal, (c) mode of attachment to tractor—mounted or trailed, (d) source of pump drive—p.t.o., auxiliary engine or hydraulic motor, (e) special-purpose machines and spraying attachments.

SYSTEMS OF LIQUID DELIVERY

Hydraulic

The diagram in Fig 126 shows the main components of a sprayer and indicates the flow of the liquid. Thus, the hydraulic principle of liquid movement by the action of a pump can be seen. This widely used type of sprayer is discussed in further detail later.

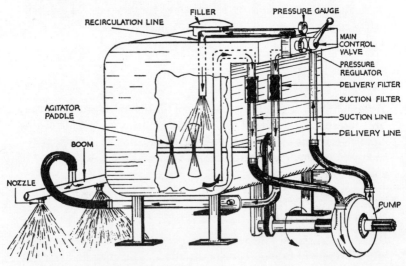

126 Hydraulic field crop sprayer

Pneumatic (pressurised tank)

The liquid delivery from this type of machine is achieved by pressurising the partially filled tank. The diagrammatic layout in Fig 127 shows an air compressor designed to force the liquid from the tank to the boom. This arrangement is particularly suitable for spraying acids and suspension type materials, as no pump

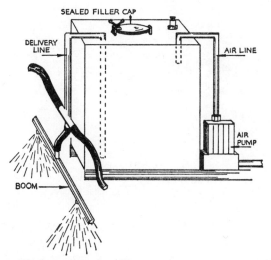

127 Pneumatic (pressurised tank) type sprayer

operates in the actual liquid which may be corrosive or abrasive. The tank of such a machine must be not only airtight but also strong enough to stand the operating pressures. Excess pressure is relieved by an adjustable relief valve. This type of sprayer is not particularly common as a farm machine but quite a number are used by contractors.

Pneumatic (air-blast atomisation)

The air-blast atomising machine qualifies as a pneumatic sprayer, although the air is employed not to discharge the liquid from the tank to the boom, but to break it down into fine droplets and to assist its distribution as it leaves the nozzles which are fed with the dilute chemical by a low-pressure hydraulic pump (Fig 128).

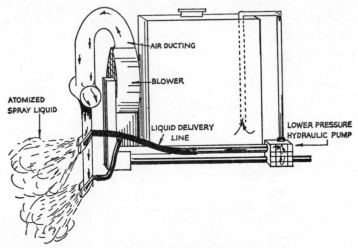

AIR DUCTING

BLOWER

ATOMIZED
SPRAY LIQUID

LIQUID DELIVERY
LINE

LOWER PRESSURE
HYDRAULIC PUMP

128 Pneumatic (air blast atomisation) type sprayer

Again, this type of sprayer is more widely used by contractors than by farmers. The air-blast nozzles are claimed to be particularly effective for the application of insecticides and fungicides, where complete coverage is so important.

VOLUMETRIC CAPACITY

Application rates are sometimes designated as follows:
high volume — exceeding 60 gallons per acre
medium volume — 20–60 gallons per acre
low volume — 5–20 gallons per acre
very low volume — less than 5 gallons per acre
With high-volume spraying, the chemical used is heavily diluted and applied as a relatively coarse spray to saturate the weeds or crop foliage. This may be necessary because of the nature of the chemical, or the object of the treatment. The *run off* of the liquid

from the crop foliage can provide a safety factor, as excessive quantities of the chemical are not retained long enough to cause damage. Some treatments can be given at either high or low volume. In a dense crop, high-volume spraying helps the liquid to penetrate through the crop foliage to the weeds.

Spraying machines designed specifically for high-volume rates of application are generally larger and more expensive, although some universal-type machines, available at lower cost, are satisfactory for occasional high-volume applications.

High-volume spraying entails the supply and transport of large quantities of water, and the weight of the outfit can cause heavy wheel marks on growing crops. However, there is no satisfactory alternative to high-volume spraying for certain classes of work.

With low-volume spraying, the smaller quantities of liquid used are necessarily more concentrated. Finer spray droplets are necessary to achieve adequate and even distribution. The low cost, small water requirements, and simplicity of operation of low-volume sprayers, are attractive features and largely account for their widespread use. The inherent problem of drift with low-volume spraying, that is the ultra-fine droplets of liquid carrying in the air to susceptible crops in the vicinity, has been reduced by special nozzles and booms which have been designed to reduce atomisation of the liquid without impairing its covering power.

Some sprayers are limited to a narrow range of application rates. Others are more versatile, and are capable of spraying at a wide range of application rates. One encounters such terms as *low volume, medium/low volume, high volume, high/low volume* and *universal* used by manufacturers in describing their machines.

The principal factors which influence the capacity of a sprayer are the size of its tank, the size and type of pump employed, and the size, number and type of nozzles used. Also, as some spray substances for high-volume application are suspension type, most high volume and universal machines have a mechanical agitating device incorporated in the tank. Other machines rely upon the re-circulation of liquid discharged from the pressure-relief valve to keep the contents of the tank stirred.

MODE OF ATTACHMENT TO THE TRACTOR

Most sprayers used on the farm are of the mounted type. Many of them are attached to the three-point linkage system, which is either locked or otherwise supported to relieve the hydraulic system of the weight of the filled tank. This direct attachment of the sprayer eliminates the second pair of wheel marks caused by trailed machines.

Some machines designed specifically for high-volume spraying are trailed, because of the large-capacity tank usually fitted to such sprayers.

SOURCE OF PUMP DRIVE

The power for driving the pump of most sprayers is provided by the tractor p.t.o. On some high-volume sprayers—particularly the trailed type—auxiliary petrol engines are used. A third alternative used by some manufacturers is the hydraulic motor operated by the tractor hydraulic system.

SPECIAL-PURPOSE MACHINES

There is a variety of special-purpose sprayers made for horticultural applications and fruit-tree spraying, and many attachments are available for the spraying of row crops. One field of development is the simultaneous spraying and drilling of sugar beet. Various layouts have been designed for this, and in some cases the sprayer tank is carried on the tractor while the nozzles are secured to the toolbar or seeder units.

Construction and working principles of hydraulic sprayers

The hydraulic sprayer has been mentioned briefly above, but its wide use justifies a closer study of its main components and their working principles. Reference to Fig 126 will simplify the identification of each part.

TANK

The tank may be of steel, either galvanised or otherwise protected from the corrosive effects of the chemicals used for spraying. Even galvanised coating is not entirely impervious to some of the stronger chemicals used. A recent development is the introduction of glass fibre as a tank material, and this has proved satisfactory for most spray substances. Other common features of the sprayer tank are a large filler cap, fine mesh straining gauze, and a drain plug to facilitate drainage and flushing of the tank between spraying operations and at the end of the season's work. On some universal and high-volume sprayers, a mechanical agitator and its drive mechanism is embodied in the structure of the tank. Tank capacities of mounted sprayers range from 40 to 100 gallons. Some machines are designed so that the pump may be used to fill the tank when mains water or a gravity feed is not available. A useful accessory is an external gauge on the tank to indicate the level of its contents.

PUMP

The pump is a vital unit in the crop sprayer. Its components are invariably made of non-corrosive metals, nylon or other synthetic materials. Its output and pressure capacities are important factors in the performance and versatility of the machine. The types of pump most commonly used are: piston, gear and roller-vane. See Figs 129-31. Some sprayers employ a centrifugal pump but this cannot be used over such a wide range of pressures as the others mentioned.

Piston pump

Piston pumps (Fig 129) are favoured where pressures in excess of 150 lb/in^2 (psi) are likely to be required. They normally give long service, and are not adversely affected by the hard fine granular particles present in some suspension-type spray materials. The air-cell incorporated in the circuit on the pressure side of a pump

G   175 is to even out the pulses of pressure created by the action of the piston.

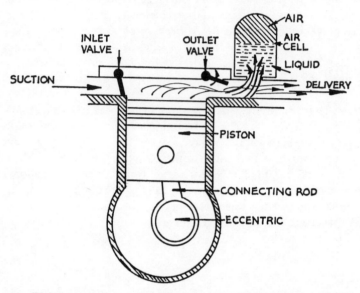

129 Piston type pump. Air cell evens out effect of piston pulses

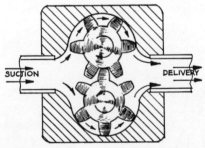

130 Gear type pump

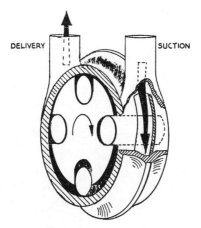

DELIVERY SUCTION

131 Roller vane type pump

Gear pump

The gear pump, Fig 130, is a small and compact unit widely used for pressures up to 75 lb/in², (psi) although some heavy-duty types are capable of much higher operating pressures. The nature of its construction, however—it has two gears in constant mesh—makes it unsuitable for use with abrasive suspension type chemicals. It may be damaged if it is operated dry, i.e. before the tank has been filled. Once filled, however, the danger of its drying up is eliminated on most sprayers by the provision of a reserve supply which is not exhausted even when the tank appears to be empty. The reserve supply of liquid is retained and simply re-circulates.

Roller vane pump

The roller-vane pump, Fig 131, has displaced the gear type on many models of crop sprayer. They have the advantage of low torque, and so require less power than other types. Various designs exist, and many are capable of operating at pressures as high as 200 lb/in² (psi). Some roller-vane pumps, however, are susceptible to damage by suspension type chemicals.

177

PRESSURE RELIEF VALVE

The pressure at which the liquid is delivered can determine the application rate, the spray pattern, the distribution and the atomisation of the spray liquid. The relief valve is situated on the pressure (or discharge) side of the pump. For simplicity, the diagram in Fig 132 shows the valve assembly as a separate unit, but the relief valve is often incorporated in the pump body itself.

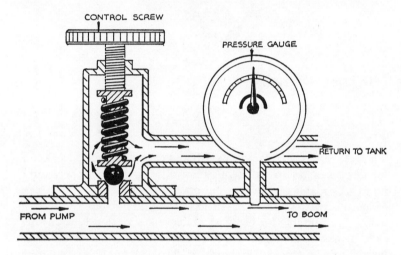

132 Pressure relief valve

The relief valve has two functions: to keep the pressure at the boom constant in spite of variations of pump speed resulting from fluctuations in tractor engine speed, and to provide a means of pre-setting the pressure at the boom to determine the rate of application and degree of atomisation of the liquid. The operator adjusts the valve by varying the tension on the spring with a control screw while the sprayer pump is running. Water may be used when making this setting prior to the spraying operation. As the adjustment is made, the effect is seen on the pressure gauge. Once the valve has been set, the excess pressure is relieved by the liquid forcing the ball off its seating and re-circulating to the tank. On some sprayers, this re-entry to the tank is used to agitate the contents. In such cases, the pump fitted is capable of a much higher output pressure than is likely to be required at the boom.

On some high/low volume (universal type) sprayers, two pressure-relief valves are incorporated in the flow circuit. The changeover from low-pressure to high-pressure operation is achieved by merely isolating the low-pressure relief valve from the circuit by means of a shut-off valve.

When work is in progress, the pressure gauge should be read frequently.

BOOM

The boom is sometimes referred to as the spray bar. It may be a metal or plastic tube, which projects symmetrically on either side of the sprayer at right angles to the direction of travel, Fig 133.

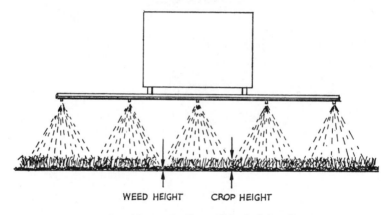

WEED HEIGHT   CROP HEIGHT

133 Correct height of boom is dictated by height of target—whether weeds or crop itself

It is drilled and tapped at intervals (usually of 18 in) to accommodate the spray nozzles. Extra tappings may be provided for a close spacing, or blanking screws for a wider spacing, for special operations.

179

For row-crop work, fan-type nozzles can be made to give a band-spraying effect by rotating them through 90° so that the flat spray pattern is in line with the direction of travel.

The boom may comprise two or three sections, and the projecting portions at each side are universally hinged at their inner ends, the liquid being supplied to each section separately through flexible hoses. The hinges have two uses: the outer boom sections may be folded upwards to reduce transport width, Fig 134, and the horizontal hinging action allows the boom to swing back at its outer end if it strikes an obstruction when in work, Fig 135. A spring returns the section to its normal position.

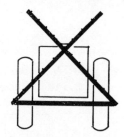

134 Boom folded for transport

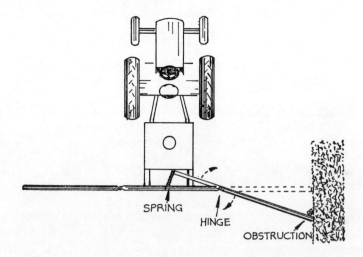

135 Hinged outer section of boom yielding to obstruction

Various boom widths are used on different machines, although most range between 18 and 30 ft. Boom extensions can be obtained for some machines to give the sprayer an increased working width, so reducing the number of wheel tracks on the crop. When these are used, the increased spraying width must be taken into account in making calibration calculations, and ground speed may have to be reduced to prevent the boom whipping if the land is hard or uneven. The slower speed will in its turn affect the application rate and further adjustments may be necessary. Another innovation is the attachment of the boom to a fore-loader to simplify height control and improve visibility.

NOZZLES

The function of the spray nozzle is to break down and distribute the liquid evenly over the target, whether this be the soil, weeds, or pests on the growing crop itself. Research to these ends has produced a wide range of nozzle types, some of which are of integral construction, while others comprise several components. The nozzles are made of non-corrosive metal, or ceramics, or of a combination of the two. Most nozzles can be broadly classified as either *fan* type or *swirl* type.

Fan nozzle

The fan type produces a flat triangular spray pattern (see Fig 136). It is used for low-volume spraying, especially where good pene-

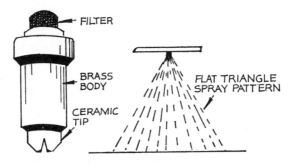

136 Brass and ceramic fan type nozzle

tration of the growing crop is required. Special wide-angle versions are available to facilitate lower boom settings when necessary.

Swirl nozzle

The swirl type produces a conical spray pattern (see Fig 137), and by selection of a suitable combination of its swirl plate and aperture disc can achieve intense or less intense atomisation; thus it is extremely versatile in its applications. The hollow-cone spray delivery results from the swirling action imparted to the liquid by angular ports or slots in the swirl plate shown. Its

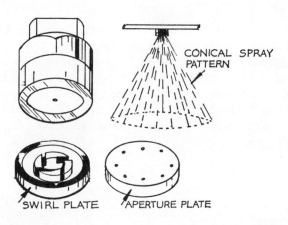

137 Nozzle types

efficient atomisation and well-defined pattern permits close range spraying where the chemical must be accurately applied. Much research has gone into sprayer-nozzle design to achieve good crop coverage without the problem of drift. One manufacturer has produced an electrically-operated nozzle which breaks down the liquid (fed at low pressure) by oscillating at high speed with a semi-rotary action, Fig 138. Adequate atomisation is achieved without the production of the very fine droplets which cause drift with the conventional design of nozzle using higher pressure.

Another manufacturer introduced a vibrating boom which reciprocates at high speed to the same end.

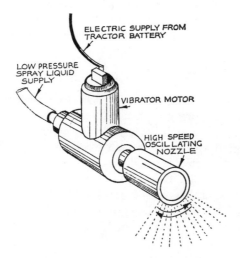

138 Electrically powered oscillating anti-mist nozzle

FILTERS

The liquid in the sprayer flow circuit must be well filtered to protect the pump from abrasion and the relief valve and nozzles from obstruction. Filters are therefore situated on the suction side of the pump, i.e. between the tank and the pump unit, on the pressure side of the pump, i.e. between the pump and the boom, and in some cases on each individual nozzle, Fig 126. A large fine gauze filter is fitted beneath the tank filler cap to filter water and/or chemical concentrate as the tank is being filled. This is particularly important when water is taken from a source other than from a mains supply. Many operators do not appreciate how easily some nozzles may be blocked by the finest particles of dirt. Even when they are not completely blocked, obstructions may distort the spray pattern to cause 'misses' in the work. Various types of filter are used, but most are constructed of fine mesh gauze and are washable. The procedure for removal and replacement is

183

usually very simple, and should be carried out frequently as a preventive rather than a corrective measure.

Blocked suction line and pressure line filters are usually revealed by a fall-off in operating pressure. Blockage of the nozzle filters has the obvious effect of reducing or distorting the spray patterns.

ANTI-DRIP SYSTEM

The anti-drip or *suck-back* system has been devised to prevent spray chemical dripping from the nozzles when the main valve is closed at the end of each bout, or during an emergency mid-field stop, as large droplets may scorch the crop foliage. The control for this anti-drip effect is usually incorporated in the main control valve. When the valve is set in the appropriate position, the flow circuit of the sprayer system is reversed so that residual liquid is sucked back from the boom and returned to the tank. Provision is generally made to prevent the pump from running dry.

Attachment of sprayer to tractor and preparation for work

When attaching a sprayer with a separate pump designed for fitting directly on to the tractor, it is usual to fit the pump first. Before fitting, however, its freedom to turn should be checked by hand. Methods of securing the pump to the tractor vary, although it is usually a simple matter if the manufacturer's instructions are carried out. The sprayer is usually carried on the three-point linkage, and a supporting stop or bracket is provided to relieve the hydraulic system of its weight. The length of the top link should be adjusted so that the sprayer tank is vertical when viewed from the side. The right-hand levelling screw should be adjusted so that the tank is vertical and the boom level when viewed from the rear (with the tractor standing on level ground). The sprayer must be rigidly secured to prevent side-sway on the linkage, and stabiliser brackets are usually fitted. With the main body of the sprayer in position, all that remains is to fit the boom sections and couple the hoses.

With these operations completed, and the tank partly filled with water, the pump can be put into operation, first with the outer nozzles of the boom removed, then with all the nozzles removed, and finally with all the nozzles replaced so that the flow circuit may be thoroughly flushed. The pump should *not* be operated without water in the tank. Whilst the machine is being run with all the nozzles fitted, a check can be made on the spray patterns produced, and the hose couplings may be checked for leaks. This also provides an opportunity to check for drift.

The sprayer should be flushed out on waste land, or some other place where traces of the toxic chemicals are not likely to contaminate pools or get into water supplies.

Before the tank is filled with the dilute chemical, the sprayer must be accurately set at the correct spraying rate, and with the right degree of atomisation. In order to achieve this, the manufacturer's data should be consulted, and it may sometimes be advisable to check the calibration of the sprayer. One method of doing this is described later in the chapter.

Operational adjustments

The adjustments described are confined to the more common hydraulic type of sprayer.

Main control valve

The main control valve regulates delivery of the liquid from the pump to the boom. In some cases, the control lever has three positions: *on, anti-drip* and *off.* The purpose of the anti-drip provision has been explained above.

Adjustment of pressure

The location of the pressure-relief valve varies considerably. Sometimes a simple hand screw is provided in a convenient position on the sprayer itself, on other machines the adjustment is in the form of a setscrew and locknut on the sprayer pump body. Regulation of the pressure is always a basic setting for the particular operation on hand, and not an adjustment to be frequently varied. The setting should comply with the maker's recommendations for the size and type of nozzle fitted. A slight divergence from the maker's basic guide may be permissible for special conditions, but excessive pressures could result in too fine a spray, with the droplets having insufficient weight to carry to the target, and the possibility of excessive drift. Insufficient pressure, on the other hand, may produce uneven liquid distribution and poor coverage.

High/low pressure changeover valve

A changeover valve is provided on universal machines to switch from high-range to low-range pressures of operation. As a rule, the control works simply by isolating a low-pressure relief valve when high pressures are required (as is normally necessary for high-volume applications), and bringing it into operation for low-pressure work.

Tank-filling valve

Some sprayers have a valve to enable the tank to be filled by means of the sprayer pump. Whenever this method of filling is used, the suction filter provided must be in place on the suction hose, especially if water is drawn from rivers or streams. The turbulence of the water entering the tank helps mixing if the chemical is poured in simultaneously.

Nozzle size and ground speed

The nozzle size and the ground speed may be included as operational adjustments as they are important factors in the performance of the sprayer.

On most machines, the liquid discharge is varied by means of either interchangeable nozzles, or nozzles having interchangeable components.

As the rate of delivery from the nozzles is constant, it is obvious that the ground speed will affect the machine's application rate.

Calibration of the sprayer

The rate of spray liquid application is regulated by three factors—operating pressure, size of nozzle aperture, and ground speed. Most sprayer manufacturers provide tables setting out the various combinations which may be used for particular application rates. This information is usually most accurate, and can be relied upon when a machine is new. After some seasons of use, however, the calibration should be checked against the maker's data. This is especially important if the sprayer is to be used for a very delicate operation where the application rate is critical. Some types of nozzles wear rapidly, so affecting the calibration. Instruction books frequently outline a suitable procedure for this check, but if such information is not available, the following sequence may be used.

1. Measure the effective spraying width of the boom, that is the width of the boom between the outermost nozzles plus the spread of the two end spray patterns.

2. Divide this into 4,356 (the number of square feet in one-tenth of an acre) to ascertain the distance the outfit will have to travel in order to spray that area. For example, if the boom width is 20 ft, the distance to be travelled is $\frac{4,356}{20} = 217.8$ ft or approximately 72 yd.

3. Measure and mark this distance on an area of land—preferably land which does not have to be subsequently sprayed.

4. Partly fill sprayer tank with water and check the level with a calibrated dip stick. Fit appropriate nozzles, and adjust pressure as recommended for required application rate.

5. Spray water over the measured course at specified speed—in this connection a low-range tractor speedometer is invaluable—and then check the quantity used by again measuring the contents of the tank with a dip stick.

6. Multiply the quantity used by ten to ascertain rate of application per acre.

7. If the quantity used does not agree with the manufacturer's table, one or two further checks may be necessary after altering

any one of the factors which affect application rate. Slight increase or decrease of ground speed sometimes proves to be the most convenient.

Boom height

The optimum height for the boom depends on the level of the *target*, and the spray patterns produced by the nozzles. Ideally, the bases of the spray patterns should just meet at the level of the target. However, as the boom height is bound to fluctuate when the outfit is in use, it is usually better to let the patterns slightly overlap rather than to risk missing strips (see Fig 133).

For critical spraying operations some makers make special recommendations for boom-height settings (i.e. nozzle to target) and quote specific measurements for different nozzle/pressure combinations. The optimum height is dictated by the nozzle characteristics.

Special attachments

For greater versatility, various attachments are available for different machines; some of the more common ones are:

1. Blanking-off screws for nozzle inserts. These are used when nozzles are to be set at specific spacings.

2. Drop legs. These are fitted to the boom as tubular vertical extensions on its underside. At their lower end, one or two orifices are angled upwards for spraying the underside of the crop foliage (Fig 139). Some makers offer an alternative type of drop leg for band spraying in row crops. In this case, the extension tube is cranked or *swan-necked* so that the nozzle can be adjusted laterally in relation to the crop rows.

3. Alternative relief-valve springs. These are sometimes made available by manufacturers to give a greater range of operating pressures, but they should only be fitted as and when recommended.

4. Extensions for boom to increase the operating width and so reduce wheel marks.

5. Hand lance. This attachment is useful for spraying small areas which are inaccessible to the tractor-mounted sprayer. Extension hoses are necessary, and care must be taken not to overdose the crop or plot on which it is used.

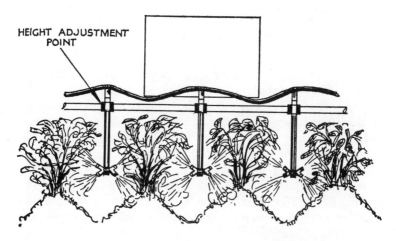

HEIGHT ADJUSTMENT POINT

139 Drop leg attachments for spraying underside of foliage

6. Water gun. This is a useful attachment for washing down machinery, yards, and walls and floors of buildings, especially when used on sprayers having powerful pumps with high-output pressures.
7. Alternative types of nozzles. In view of the differing characteristics of fan-type and swirl-type nozzles it is as well to keep both on hand.

Field procedure

TIMING OF THE SPRAYING OPERATION

Spraying must be done at the right time. This cannot be over-emphasised. It must be done when weeds and pests are most susceptible to the treatment, and also before the crop is so far advanced that it will be damaged by the chemical applied or by the tractor wheels. Excessively soft soil conditions can result in wheel-mark damage from which the crop may never completely recover.

The control of weeds in an undersown crop needs special care in timing and selection of the chemicals.

FILLING THE TANK AND MIXING THE CHEMICALS

Instructions for diluting and mixing spray chemicals are normally given on the chemical container. Some substances require pre-mixing, while others do not. The suspension type materials usually need vigorous agitation, both during the filling process and whilst spraying is in progress. After any necessary pre-mixing, the chemical should not be poured into the tank until the latter is about one-third full. If the pump circuit can be isolated from the boom, the pump should be run to agitate the liquid while the tank is being filled. The agitation should continue for several minutes after the tank has been filled, and also, if possible, whilst travelling to the field. When sulphuric acid is used, water should be put in the tank first and the acid added to it. It is extremely dangerous to add water to acid. Chemicals should never be stored where animals can reach them, and used containers should be destroyed or disposed of so that they cannot harm animals, birds, or bees.

WEATHER CONDITIONS FOR SPRAYING

When spraying, the atmosphere should ideally be humid but without rain, warm but without strong sunshine, and windless. The occasions when all these requirements coincide are rare indeed, but spraying often has to go on. One condition, which must be avoided at all costs, is high wind, or even a strong breeze if susceptible crops are in the vicinity. The 'vicinity' is difficult to define, as drifting chemical has been known to cause damage half a mile away from the actual operation. The work should not be attempted if the foliage of the crop is wet, or if rain appears to be imminent.

SEQUENCE OF SPRAYING

The headland should be sprayed *first*. Inevitably, some crop foliage is damaged by the tractor wheels, especially if it is well developed. The crop usually recovers from this physical damage, but the application of some spray chemicals to the crop *after* repeated bruising or crushing allows the chemical to penetrate the plant system, and this can have adverse effects. When the headland has been sprayed, the general direction of the cross-field work has to be decided. It is usual, to save fuel and time, to travel parallel to the longest axis of the field, although it may be preferable to work *across* the wind direction starting at the leeward side of the field. This reduces the unpleasantness for the operator normally experienced with a following wind. It is especially important when using poisonous chemicals.

DRIVING TECHNIQUE

When spraying, it is not easy to see the boundary of the completed area, and it is difficult to match up the spray bouts accurately, particularly for the novice. Until good judgment has been acquired, it is advisable to use markers, stopping at each bout end and stepping out the required distance for the next run. When spraying a cereal crop, it is sometimes convenient to follow the drills to keep an accurate course. Sometimes, however, spraying across the drills is recommended to obtain better foliage penetration when applying insecticides.

The influence of ground speed on application rates, and on boom-height variations due to excessive whipping, must not be forgotten. For critical work, particularly on hilly land, the tractor engine governor must be efficient, or discrepancies in application rates will result. It is important to reduce speed when turning at the bout ends, to avoid unnecessary wheel damage to the growing crop.

To negotiate the narrow headlands of potato fields, some sprayers have a control lever by which the operator can raise the boom sections vertically. The control valve must be opened and closed accurately at the end of each bout to avoid both missing and overlapping.

REFILLING THE TANK

In order to avoid running out of spray liquid in mid-field, some operators refill the tank well before it is empty. This practice is not as wise as it may appear, because unless the tank is refilled with a prepared mix of the correct proportions, the dilution of the concentrate is affected. It is safer to use up almost the entire tank contents before refilling. It is in this connection that an external visual tank-contents gauge is a great asset. If the outfit runs out of liquid in mid-field, the operator should pull over on to the strip last sprayed to make the 'light' run. This is recommended for the same reason as is the initial spraying of the field headlands, that is to avoid bruising the crop *before* spraying.

OPERATIONAL CHECKS

During the spraying operation, close watch should be kept on the following:

1. Pressure gauge. This is necessary to ensure that the required value is being maintained. Reduction in pressure could be due to a shortage of liquid in the tank, a blocked filter, an air leakage into the suction line between the tank and the pump, an obstructed relief valve, or a leaking hose.

2. Nozzles. These should be watched for any deformity or inconsistency of spray patterns, and of course for complete blockage. Spare nozzles should always be carried, so that a defective nozzle can be replaced at the end of a bout or on the spot if the sprayer has anti-drip control. The blocked nozzle can later be cleared by applying an air line or pressure water jet at the discharge end of the nozzle—do not probe the fine orifice with wire.

3. Degree of drift. The dangers of spray drift have been mentioned. Even though the atmosphere may be windless when spraying begins, a wind may well develop before the work has been completed. If it becomes strong, it may be possible (after reference

to the maker's data) to continue working by fitting larger aperture nozzles and using a lower operating pressure to increase the droplet size and reduce drift. A change of nozzles and pressures usually produces a slightly different spray pattern, necessitating re-setting of the boom height. Sometimes drift can be reduced by weakening the concentration of the liquid and increasing the rate of application.

4. Band spraying. This requires accurate nozzle height which inevitably affects the width of band and consequently the concentration of chemical near the seed.

PROCEDURE WHEN CHANGING CHEMICALS

When changing from one chemical to another between different spraying operations, it is absolutely essential that the tank and the whole sprayer system are thoroughly cleared of the first substance. To carry out this operation, the sprayer should be taken to an area where the contaminated water, discharged from the machine during the flushing process, is not likely to contaminate any water courses or remain in puddles on the surface as a hazard to live-stock and wild life. Some sprayers always retain a certain amount of liquid to keep the pump flooded, so the hose in that circuit should be disconnected, or the drain cock opened, so that this liquid may be extracted. At least 30 gallons of water should be pumped through the sprayer system, and all filters and nozzles should be removed when half of this amount has been put through. The gauze at the filler orifice should be thoroughly washed and any spilt chemical on the outside of the tank washed off and brushed away.

Special detergents are available for the decontamination of sprayers, although soda is quite effective when used in a ratio of 1 lb to 10 gallons of water. When such cleaning solutions are used, it is better to repeat the process two or three times with small quantities of liquid than do it once with a large quantity. Certain ester formulations are so penetrating that the boom hoses should be removed and left to soak—fresh ones being fitted for the new operation.

193

Maintenance

DAILY ATTENTION

Any lubrication requirements are usually limited to the pump, and, where fitted, the power-drive shaft universals and the mechanical agitator. During a heavy spraying programme, the filters should be cleaned daily, especially when water is used from a source other than from the mains. When replacing the filter elements, the gasket or seal of the filter body must be in good condition. If it is split or distorted, air will be drawn in or liquid will be forced out, depending on whether the fault is on the suction or the delivery side of the pump. After each day's work, any liquid remaining in the tank should be drained off, and the whole machine thoroughly flushed out. This is particularly important when using very corrosive chemicals. This procedure might appear unnecessary if it is intended to resume work the next day, but inclement weather often precludes a follow-on of work, and as a result a sprayer tank left partly filled for several days becomes heavily contaminated with the chemical.

END-OF-SEASON ATTENTION

The sprayer must, of course, be thoroughly cleaned at the end of the season.

As it is a relatively simple machine having few moving parts, end-of-season attention involves little mechanical work. All filters should be removed and inspected for damage, as should the pressure-relief valve and its spring. Some types of nozzles or nozzle-tips do tend to wear rather quickly, and so the complete set of nozzles should be changed as recommended by the maker of the machine. After several seasons of service, the pump of the sprayer may need reconditioning or replacement due to wear on its moving parts. According to the type of pump fitted, such parts may include bearings, gears, rollers, pistons or valves, although overhaul of the pump is best left to the agricultural engineer. Pressure gauges sometimes become sluggish in action, or fail to return to zero; in such cases these should be replaced. A suspect pressure gauge is usually tested by comparison with a new one. Hoses should be checked, and the protective coating or lining of the tank should also be inspected. Finally, the machine should be greased if and where nipples are provided, and the exterior of the sprayer should be protected against corrosion.

REPLACEMENT PARTS

Items under this heading include nozzles, filters and boom and pump hoses.

Safety precautions

Most farm machines require careful handling and an awareness of danger points, but spraying machines can be particularly hazardous. Through flagrant disregard of warnings, many operators have suffered serious poisoning. Even when some so-called non-poisonous substances are used, inhalation of large quantities of the spray mist can cause discomfort. Before using any spray material, operators should ascertain whether or not the substance is scheduled as a poison. Publications are issued periodically listing newly introduced toxic substances. Moreover, statutory regulations enforce positive precautionary measures when such chemicals are used. Unfortunately, some of the poisons can enter the human system in a number of ways—by inhaling, by contact with the mouth or stomach through eating or smoking, or even more insidiously through the pores of the skin. The subtleness of the action of some of these chemicals is such that no ill effects are sensed by the victim until the cumulative effect of contamination, from repeated periods of spraying, results in serious poisoning of the system. So, it is advisable for operators to acquaint themselves with the following safety precautions:

1. If the chemical is scheduled as a poisonous substance, wear protective overalls, rubber boots, gloves, eye-shields and mask. These clothes should be cleaned frequently. Operators engaged regularly in contract spraying must also have air-conditioned cabs on their tractors.

2. Follow rigidly the instructions supplied with the chemical.

Handle the substance carefully and avoid splashing. Destroy used containers.

3. Where convenient, spray at right angles to the wind direction, starting at the leeward side of the field to avoid, as far as possible, the inhalation of spray mist.

4. Refrain from eating, drinking and smoking when at work. In addition to their toxic effects, some spray substances are also highly inflammable.

5. Never put a blocked spray nozzle to the mouth to blow it clear; instead it should be cleared by one of the methods described under *Operational checks*.

6. Wash hands and face thoroughly after work and before attempting to either eat or smoke.

7. When work has finished for the day, do not leave the sprayer where children or animals can touch it.

8. If any ill-effects are experienced after a period of spraying, consult a doctor immediately and *advise him of the chemical you have been using* so that he can prescribe the most effective treatment.

6 FORAGE MACHINERY

The term *forage* may be applied generally to all kinds of animal food, but it has recently come to mean herbage cut in a more or less green state for conservation by artificial drying or by ensiling. Each of these operations sets out to preserve the material for later use, with the minimum loss of nutrients in the process. The traditional procedure of field haymaking is still the most common method of fodder conservation, but heavy losses of nutrients occur if the weather is poor or if the crop is harshly processed by swath treatment machinery or balers. The collection of crops intended for drying or ensiling is less dependent on good weather, and the conservation processes permit the crop to be cut and collected when at its richest in feeding value.

CROP DRYING

The principle of artificial crop drying is simply to reduce the moisture content of the crop to a precise value under controlled conditions. This produces very high quality material, assuming, of course, that the quality potential is in the crop initially. However, the output of farm crop-drying plants is slow by comparison with the speed of silage making. Larger installations are expensive to install and operate, and require sophisticated organisation and management. Also, the high costs of drying are justifiable only on very high quality herbage—some installations are used almost exclusively for the drying of lucerne—of course high quality material is also important for profitable silage making.

ENSILING

The principle of ensiling is to compact suitable green material so as to promote a sequence of complex chemical changes in the material (aided by the action of certain bacteria) and thereby preserve it for later use. Expert management of field and silo operations is necessary to ensure the correct degree of beneficial fermentation and acid formation to preserve the material in a nutritious and palatable form. The precise techniques adopted depend upon the nature and stage of maturity of the crop, on the end product desired, and on the equipment available.

Various types of ensiled material may be distinguished as *silage*, *drylage*, and *haylage*. The most significant distinction between these various forms of ensiled material is their moisture content. Material with a moisture content in excess of 65% may be classified as silage; material with a moisture content ranging from 50% to 65% as drylage; material with a moisture content less than 50% as haylage.

The stage of maturity at which crops are cut and collected influences the dry matter/protein ratio.

The merits of the various types of ensiled material will not be debated here, but reference will be made to the influence that machine design and operation can have on them. For instance, if the cutting and collection can be carried out separately, wilting can occur between the two processes, which will reduce the loss of nutrients and silage effluent. The two operations entail the use of either two separate machines or one machine which can be adapted for each operation. Another factor which can influence the quality of the ultimate silage is the uniformity of compaction, and it is useful if a machine can lacerate the crop, particularly when conserving long and coarse materials.

As the various machines are considered, it will be seen that

198

fine chopping of the material has further advantages for certain applications.

Construction and operation of buckrakes

Now that the potentialities of green-crop conservation are better understood, a wide variety of crops are grown expressly for ensiling. Also, some crops are conserved in this way as a matter of expediency. There are various methods and systems of mechanising crop conservation and these have led to the introduction of different types of equipment such as buckrakes, pick-up loaders, cutter loaders, chopper-blower harvesters, flail-type harvesters and flail-chop harvesters. These machines are of varying versatility, although the two last have by far the greatest range of application. Pick-up loaders and cutter loaders are now obsolete so reference to these will be omitted.

PRINCIPLES

On some small farms silage is made with tractor-mounted buckrakes as the sole means of collecting crops which have been previously cut with a mowing machine. As shown in Fig 140, the buckrake is a simple attachment mounted at the rear of the tractor on the three-point linkage. It is relatively cheap to buy, and it finds many other applications apart from silage making. The mower and buckrake system of green-crop collection permits a period of wilting between the two operations, which is advantageous in the making of silage with some types of herbage. When filling silage clamps, the tractor and rake can be driven over the material to discharge and help spread the load. Its passage over the clamp helps compact the material to prevent excessive heating during the subsequent curing period (since oxygen accelerates the heating process).

The relatively small carrying capacity (usually between 3 and 7 cwt) of the buckrake makes it uneconomical for hauls much in excess of 500 yd. Although the buckrake is a relatively simple attachment to use on firm level land, considerable skill is necessary in fields which have pronounced ridges and furrows, especially if the surface of the ground is soft.

The buckrake is sometimes used with other green-crop loaders and forage harvesters, to convey the collected crop from the trailer-unloading area to the clamp. For this work, thin solid steel tines penetrate short material better than the tubular type.

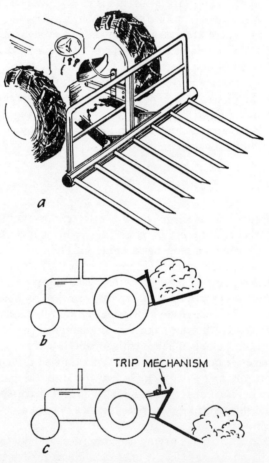

140 Tractor and buckrake: *a* Mode of attachment, *b* Loaded rake, *c* Rake released for unloading

ATTACHMENT TO TRACTOR AND PREPARATION FOR WORK

Actual attachment of the buckrake to the tractor entails its connection to the three-point linkage. When this has been done, the adjustable lift rod of the tractor linkage should be set so that the buckrake is parallel with the ground when viewed from the rear. The top-link length should be set so that the pitch of the tines, when lowered to their operational position, is just sufficient to allow their points to float on the ground surface without digging in, although this adjustment may have to be modified in the field according to conditions.

Before commencing work, the operator should check if a device is provided on the tractor to relieve the hydraulic system of the weight of the loaded buckrake. If it is not, the operator must fully understand how to use the tractor hydraulic control levers. The only operational control on the rake itself is the trip-release mechanism provided for discharging the load (Fig 140c). Any adjustment provided on this is usually a maintenance feature rather than an operational setting. The pressure of the rear tractor tyres should be checked before starting work, as it is possible, when working in long material, to collect quite a substantial load on the rake, and this imposes considerable weight on the tyres.

OPERATION

The precise system of working in the field depends upon the type and volume of the crop to be collected. Sometimes the material may be gathered directly from undisturbed swaths, with the buckrake collecting two swaths at a time, travelling in the same direction as that of the mower. Alternatively, better loading can sometimes be achieved by working across the swaths. With windrowed or crimped swaths, a trial will show the most effective method of collection. Sometimes pronounced ridges and furrows in a field dictate the direction of working. Tractor speed also can affect the efficiency with which the buckrake picks up the material, but a test run will reveal the best speed.

A loaded buckrake, particularly when in the raised position, decreases the tractor's stability—its rear overhanging weight tends to lift the front of the tractor, so impairing its steering. The higher centre of gravity produced by the raised load reduces the tractor's lateral stability, so that driving across a hillside or turning sharply at speed can turn it over sideways. The common practice of driving the tractor on to the silage clamp, in order to discharge the load, requires skill and care. The tractor should be reversed up the ramp of material, not driven forward.

Construction and operation of chopper-blower harvesters

The importance of even and consistent compaction in making good silage has been mentioned. This is not always easy to achieve with the long-coarse and bulky materials sometimes ensiled. Compaction of the material is particularly difficult in tower silos, and it was principally for this reason that machines were introduced incorporating a chopping unit to reduce the crop to short lengths (Figs 141 and 142). Such chopped material is virtually self-compacting after even distribution. Other advantages of chopping and/or lacerating are quick and even wilting, where this is desirable, and ease of handling the resultant end product particularly if this operation also is mechanised).

Most chopper blower machines employ a pick-up, and process a previously cut crop, but some of them can be readily adapted with a reciprocating cutter bar to harvest a standing crop. An auger and elevator immediately behind the pick-up or cutter bar conveys the crop into the chopping unit. The pressure roller or upper conveyor slightly compacts the train of material, and feed rollers ensure an even flow to the chopping mechanism. This control of the feed is sometimes described as 'metered feed'. Two types of chopping mechanism have become firmly established; these are usually referred to as *cylinder type* and *flywheel type* and a brief explanation of their working principles follows.

Cylinder type chopping mechanism

Fig 141 illustrates the layout of a typical machine employing a cylinder type of chopping unit. The chopping mechanism consists

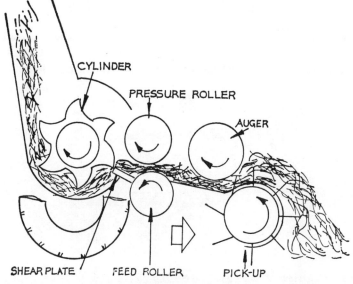

141 Forage harvester with cylinder type chopping mechanism

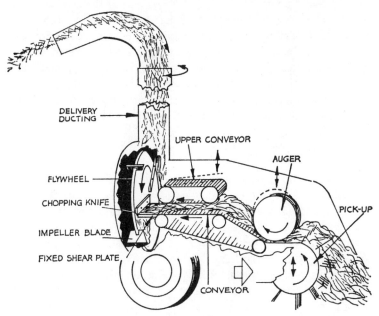

142 Forage harvester with flywheel type chopping mechanism

of a cylinder formed by a number of spiral blades which work in conjunction with a fixed blade on to which the crop is fed, in a similar manner to a lawn mower, Fig 143. The finished length of

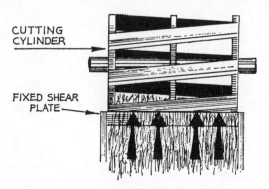

CUTTING
CYLINDER

FIXED SHEAR
PLATE

143 Plan view of cylinder type cutting mechanism

material may be regulated by the entry speed of the crop into the cylinder chamber, by the speed of the cylinder, or by both these factors. The optimum length of 'chop' depends mainly upon the type of crop being harvested and the proposed type of silo to receive it. After chopping, the material is delivered from the chamber into an accompanying trailer by a fan-type impeller unit. The positive chopping action and the mode of discharge gives this machine its distinguishing name of 'chopper-blower'.

Flywheel type chopping mechanism

In this case the chopping element consists of a heavy circular steel plate rotating at right angles to the direction of the crop intake, and fitted with a number of blades standing off its forward face (Fig 144). This rotor develops considerable momentum —hence the term *flywheel type* used to distinguish this type of chopping unit from others. The material is chopped as it passes over the edge of a fixed steel ledge. The length of 'chop' can be regulated by varying the speed of the material's entry, as with cylinder-type choppers, by varying the flywheel speed, or by varying the number of blades mounted on the flywheel, Fig 145.

The chopped crop is ejected into a trailer by impeller blades, which are generally mounted on the same flywheel as the chopping blades. Plate 10 shows a machine of this type at work.

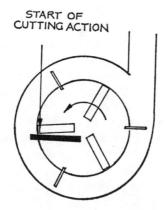

144 Shear plate positioned below flywheel centre reduces chopping impact

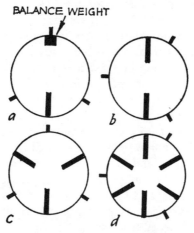

145 Arrangement of chopping knives on flywheel face: *a* For one knife, *b* Two, *c* Three, *d* Six

Construction and working principles of flail-type harvesters

Some eighteen years or so elapsed between the first appearance of the flail-type forage harvester in this country and the production by British manufacturers of similar machines on any appreciable scale. Once its efficiency and potentialities became known, however, the machine virtually revolutionised the cutting and collection of green crops. The working principles of the flail harvester may be described as a new departure from those employed in earlier types of forage equipment.

Fig 146 illustrates the main components and working principles of a typical machine. The blades, which are made of high-grade steel, are attached to a rotor at one end, allowing the free sharpened end to swing with a flailing action, Fig 147. The rotor, which is p.t.o. driven through the media of a gearbox and multi-vee belts, (Fig 148), runs at high speed in the opposite direction to the land wheels, severs the crop from the sward, and lacerates it against an adjustable shear-bar. Partly by the impelling action of the flails, and partly by the air current they create, the lacerated material is delivered up through the trunking and discharged from a spout into an accompanying trailer. Even from this cursory description the simplicity of the flail-type harvester is obvious, and this largely accounts for its relatively low cost, general reliability and absence of complicated maintenance requirements.

It is also versatile, not only for green-crop harvesting, but in numerous other operations on the farm such as kale and maize cutting; grass topping; the opening out of mowing grass around the field boundary for the mower; hay conditioning by using the harvester as an alternative to the crimper or tedder; straw and green manure chopping, shredding or spreading; the cutting down of 'bolted' sugar beet, and the pulverising of potato haulms.

So far as green crops are concerned, the machine may be used for cutting, lacerating and loading; cutting, lacerating and windrowing (where wilting is desirable), or collecting, lacerating and loading of wilted crops from swaths or windrows. In all these operations, the harvester is capable of a very high rate of work.

Although the flail harvester is simple in design, it must of

206

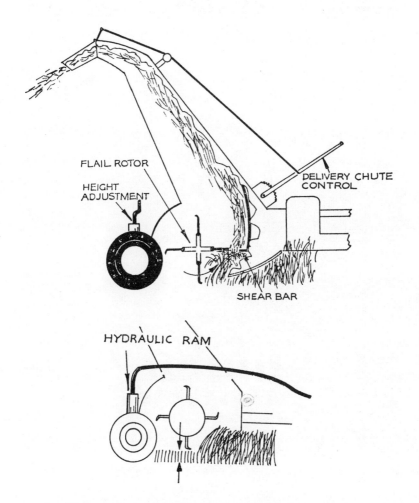

FLAIL ROTOR

HEIGHT ADJUSTMENT

DELIVERY CHUTE CONTROL

SHEAR BAR

HYDRAULIC RAM

146 *a* Flail type harvester, *b* Hydraulic cutting height control

necessity be extremely robust in construction. Also, in view of its high power requirement, it must be correctly matched to the tractor with which it is to be used.

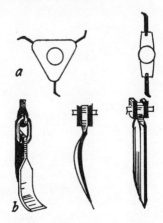

147 *a* Two- or three-bank flail rotors, *b* Alternative types of flail

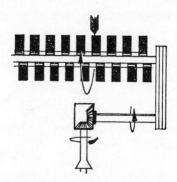

148 Rotor drive system

TYPES OF FLAIL HARVESTERS

Apart from the finer points of technical specification, and the various adaptations which can be performed on the different makes of flail harvesters, the general features of design have settled into a fairly uniform pattern. However, two important optional features offered by some makers are the incorporation of a secondary chopping unit, and alternative layouts to give a choice of operational positions in relation to the tractor.

Secondary chopping unit

Harvesters fitted with a secondary chopping mechanism are sometimes referred to as *double-chop* machines. Fig 149 illustrates the usual arrangement. First, the flail blades cut the crop, and

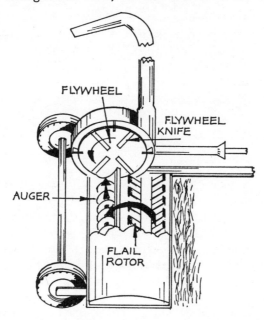

149 Double-chop forage harvester employing flails and flywheel chopper

throw it up into a trough which houses an auger. Then the auger feeds the material cross-wise towards a flywheel-type chopper unit at one side of the machine. The principles of chopping and discharging are similar to those employed on the flywheel-type chopper-blower machines described previously.

For some work the lacerating effect of the flails is adequate, but the secondary chopping action is advantageous when handling long and bulky crops, as it simplifies the spreading and compaction of the material at the silo. As mentioned previously, fine cutting is an advantage, and may be a necessity where tower silos are to be used, especially if the discharge of the material from the silo, and the distribution to the stock, are mechanised.

Alternative layouts

Flail-type forage harvesters are currently classified by manufacturers as *in-line, offset* and *side-attached*. These terms describe the position of the flail rotor in relation to the tractor. However, to avoid confusion between the two last, they will be referred to here as *in-line, offset (rear-attached)* and *offset (side-attached)*. (See Fig 150 and Plate 9 which shows an offset machine at work.)

In-line

The in-line type of machine for any given operating width is generally cheaper than either of the other types. Its position relative to the tractor, simplifies the negotiation of gateways and narrow lanes, Fig 150a. This is an asset when working with a single tractor, where the whole outfit—tractor, harvester and trailer—shuttle from field to farm. This system is quite often preferred when collecting green crops with a small labour force and for the occasional topping-off of irregularly grazed pastures. (Some offset machines can be moved more in-line for transport, but this requires a little time and effort.) The in-line harvester is also free from the side draught which exists with some offset machines, although this is not, as a rule, particularly troublesome.

One disadvantage of the in-line machine is that at least one pair of tractor wheels must run on the uncut crop. To reduce the effect of this, the tractor wheels should be set wide enough to clear the cutting width of the harvester, and the outfit should be driven in opposite directions on adjacent bouts across, or circuits around, the field so that the material flattened by the wheels in one direction may be retrieved on the adjacent bout, Fig 150a. In many cases, it is possible by these measures to retrieve all the flattened material. Problems could arise when picking up very long material or wilted crops which have been flattened by the tractor wheels. The wheels sometimes soil the material when collecting green cereals or similar crops from arable land, especially if the soil is wet.

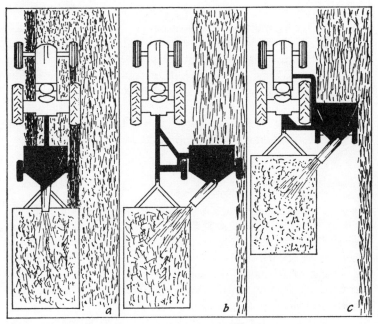

150 Flail harvester layouts: *a* In-line, *b* Offset (rear attached), *c* Offset (side attached)

Most in-line harvesters are fitted with a rear drawbar hitch point for the attachment of a following trailer, if required as an alternative to an accompanying tractor and trailer outfit. However, this may give rise to tractor wheel-spin, especially on hilly land and on wet surfaces, as none of the trailer's weight is transferred on to the rear of the tractor. Some harvester manufacturers have placed the harvester's land wheels well back on the chassis to transfer some of the harvester's weight on to the tractor, Fig 151. When working with the outfit in a tractor-harvester-trailer train, visibility of the loading from the tractor driver's seat is somewhat restricted.

Offset (rear-attached)

The offset (rear-attached) machine avoids the problems of wheel marks and gives the tractor driver better visibility, both of the crop in front of the harvester, and the discharge into the trailer, Fig 150b. A swinging drawbar attachment is usually provided to reduce the width of the outfit for transport. As with the in-line machine, however, when the harvester is 'in train' with the tractor and trailer, the weight of the latter is carried on the rear of the harvester and not on the tractor drawbar, so that traction can still be a problem on hilly land and in greasy conditions.

Offset (side-attached)

Like the rear-attached, the side-attached type also avoids the problem of wheel marks in the uncut crop and permits good all round visibility from the tractor. The fact that the rear of the tractor is left clear for the direct attachment of the trailer, allows the weight of the latter to be transposed on to the rear of the tractor so improving wheel adhesion, increasing manœuvrability by eliminating the lag at corners, and simplifying coupling and uncoupling of trailers (Fig 150c). In fact, the attachment and detachment of some side-attached harvesters is so simple that, if continuous cutting is not required, the harvester may be quickly detached to allow the tractor with its attached trailer to transport the material to the silo. A further advantage of this abreast position of the harvester is the extra stability it provides for working on hillside land.

Despite its secure attachment to the side of the tractor, vertical flexibility of the harvester is preserved so that it may follow the contours of the land. For transport, the machine may be trailed in line with the tractor.

MECHANICAL PROTECTIVE DEVICES

The flails have some protection from impact with obstructions, as they are free to swing on the pins and bushes which attach them to the rotor. In many cases on simple flail harvesters, no further protection is incorporated in the drive than that provided by the incidental slippage of the belts when the rotor is overloaded. Slippage under normal working conditions, however, should always be avoided by ensuring that the belt is at the recommended tension.

Double-chop type machines usually have a friction safety clutch incorporated in the drive system.

Over-run clutches are also fitted on some machines. These operate rather like a ratchet in the power-drive system, relieving the latter of the momentum of the harvester rotor when the p.t.o. drive is declutched or disengaged. This allows a quicker reduction of the speed of the power-drive shaft when the latter is momentarily disengaged for cornering or gear-changing. Some makers offer the over-run clutch as an optional extra.

Another safety factor sometimes employed in double-chop machines is a shear-pin, designed to transmit normal operating torque but to shear off in the event of overload.

ATTACHMENT OF HARVESTER TO TRACTOR AND PREPARATION FOR WORK

Earlier chapters have made it clear that careful matching of machines with the tractors which draw them is most important. The forage harvester's relatively high power demand on the tractor makes this consideration imperative.

Attachment details vary from one machine to another. The variations include required drawbar height, power-drive shaft length, and the use of special drawbar attachments or linkage stabilisers. In all these matters, the manufacturer's recommendations should be followed. In common with all power-driven machines, the power-drive shaft must be correctly aligned to provide the necessary flexibility vertically, laterally and lengthwise.

For machines with the fairly common cutting width of 40 in, a tractor developing at least 25 bhp is recommended. For machines of 60 in cutting width, 30–35 bhp is recommended. Double-chop type machines impose a greater load on the tractor, which must be taken into account. Such matters must be discussed with the suppliers before a machine is purchased. Most machines are basically designed to operate at a p.t.o. speed of 540 rev/min, and adaptors are available to match the power-drive shaft spline sizes where these are different.

With in-line harvesters, the tractor wheels should be set to straddle the machine's cutting width. Even with offset machines, it may be necessary to re-set the wheels to clear the path of the rotor. With side-mounted machines, the tyre pressures of the rear tractor wheels should be adequate to carry the extra weight imposed on the drawbar by the loaded trailer.

OPERATIONAL ADJUSTMENTS

The simplicity of the flail harvester's design has misled many uninitiated operators so far as operating technique is concerned, for although it possesses few operational adjustments it is important for efficiency and economy that these are judiciously set.

Cutting height

The basic cutting height is usually set by raising or lowering the machine on its land wheels. This is achieved either manually by means of a screw lever or turnbuckle, or hydraulically (Fig 146). Where a hydraulic lift is provided, it can be operated rapidly in an emergency to avoid big obstructions in the path of the rotor. A useful feature on some machines is the ability to move the wheels foreward or back on the chassis of the harvester. Moving them forward brings them more in line with the rotor and gives more accurate height control. Moving them rearwards transfers more of the harvester's weight on to the rear of the tractor to improve wheel adhesion, Fig 151.

As the vertical wheel adjustment on the harvester is usually individual, it is important, after modifying the setting, to check that the rotor is level, i.e. parallel to the ground surface. Sometimes indicators are provided on the wheel assemblies to simplify this check. The two side flanges of the rotor housing are fitted with skids on their undersides; this ensures that a small amount of clearance always exists between the flail tips and the ground. Their

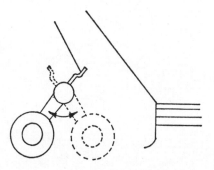

151 Alternative wheel positions. *Forward* gives more accurate height on uneven land. *Rearward* transfers more weight on to tractor wheels and improves adhesion

presence is particularly important on uneven or ridged land to prevent 'scalping' of the sward, picking up of soil, or hitting stones. A balance spring between the rotor housing and the harvester drawbar protects the machine from shock loads when working on very irregular surfaces.

Lateral wheel setting

Not all harvesters have a variable wheel setting. Adjustment is only necessary when the machine is to be used in row crops, for such purposes as cutting down potato haulms or bolted sugar-beet tops. For constant flail to ground clearance, the wheels must conform to the spacing of the ridges or rows. Adjustment is achieved by various means on different machines—the more usual being extensible axles or movable wheel mountings.

Intensity of laceration and chopping

The nature and condition of the crop itself will dictate the required laceration and chopping intensity. The severity of treatment is influenced by rotor speed, ground speed, flail/shear-bar clearance, speed and number of chopping knives (double-chop-type machines), type and sharpness of flails, and the position of the crop outlet (i.e. high level or low level).

Laceration may also be regulated by the setting of an adjustable laceration plate (Fig 152) or concave plate (Fig 153), or the fitting of a special shredding attachment. These various factors will be described in more detail.

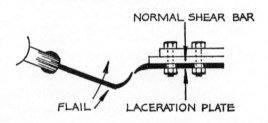

152 Laceration plate

153 Concave plate in place of normal shear bar

Rotor speed

Manufacturers' recommendations on rotor speed vary, although most harvesters have a rotor speed between 1,500 and 1,750 rev/min. In some cases, the operational speed range is specified as from 1,000 to 1,800 rev/min. A trial is usually necessary to find the

216

optimum rotor speed for any particular operation. Slight speed variation is usually possible by varying the p.t.o. speed, although this may entail a change of gear if ground speed is critical. More drastic changes may involve the interchange of vee pulleys on the rotor-drive system.

Excessively low rotor speeds reduce cutting efficiency and lacerating effect. Moreover, in heavy crops, the reduced impelling and blowing action of the rotor may cause blockages in the delivery

154 Low discharge attachment for windrowing and/or conditioning

chute and perhaps the rotor housing. To overcome this, when a low rotor speed is required (as in the conditioning of hay), some machines have a low-level delivery outlet through which the crop may escape more freely. Excessively high speeds, on the other hand, should be avoided (Fig 154), as the power demands of the flail rotor rise considerably as its speed increases, and an extra 300 rev/min may need an additional 5 or 6 hp. So rotor speed has a marked effect on tractor fuel consumption.

Ground speed

Ground speed is such an important factor when operating the forage harvester that it comes early in the list of operational adjustments. Often the crop and/or ground conditions dictate the ground speed of the outfit, but there are times when some latitude is possible and advantage should be taken of this. Speeds of up to 8 miles an hour are permissible so long as the desired effect is achieved. Where severe laceration is desirable, ground speed should be reduced by changing to a lower gear so that adequate tractor engine and p.t.o. speeds are maintained. If this is not severe enough, the rotor speed should be increased by using an alternative belt pulley ratio. For gentler treatment, the ground speed should be increased or the rotor speed decreased.

Flail/shear-bar clearance

The clearance between the tips of the fully extended flails and the shear-bar can usually only be varied between the limits of $\frac{1}{4}$ to 1 in, but even this narrow limit of adjustment has considerable effect. On some machines the shear-bar can be removed altogether to give the minimum of laceration. A special shredding plate or shear bar with knives can be attached for maximum laceration.

Chopping knives (double-chop-type machines)

The secondary chopping unit of the double-chop harvester can produce much finer material than the standard flail harvester. The fineness of 'chop' is controlled by varying the speed of the flywheel knife rotor, and/or by varying the number of knives fitted. Whenever the blades are disturbed, the flywheel must be kept in balance by arranging the blades symmetrically around the flywheel (Fig 145).

Flails

Many different types of flails are now in use, some having a forward curve sharpened only at their tips, and others cranked or curved sideways and bevelled to give a side cutting effect (Fig 147b) The former type usually cuts slightly more cleanly and gives a more consistent height, while the latter tends to have more lacerating effect, particularly with long material. A third, more

218

robust pattern is straight with a bevelled leading edge. The edge-on flail has a lower wind resistance than the broad face of the forward-curved type, thereby reducing power demand. On the other hand, strong air turbulence may be desirable for some operations—in particular when picking up a wilted crop.

To be most effective the cutting edges must be kept sharp. Some flails are double-edged and reversible, giving a double lease of life between sharpenings. After repeated sharpening, loss of flail length may affect the degree of laceration by increasing the shear-bar clearance beyond the limit of adjustment.

Crop outlet

Some harvesters have a fixed chute for the outlet of the crop, although most are now fitted with a rotating or swivelling portion at the discharge end. This simplifies discharge of the material into accompanying trailers.

Some harvesters have a low-level outlet or a hinged upper section for windrowing or for the swath treatment of hay, Fig 154. In some cases, alternative chutes are provided to replace the standard ducting. The degree of congestion which occurs in the chute of the harvester influences the turbulence inside the rotor casing and so can affect both cleanness of cutting and degree of laceration.

SPECIAL ATTACHMENTS

As the versatility of the flail-type harvester becomes more appreciated, an increasing range of special attachments are becoming available. These include the following:

1. Concave extension or adjustable front apron. These are to improve the collection of short or wilted material.
2. Shredding plate. This increases the lacerating effect of the flails.
3. Alternative types of flail. These give varying degrees of lacerating, shredding or pulverising effect.
4. Additional rotating chopping knives for double-chop type

harvesters. These may be fitted where extra-fine chopping is required.

5. Maize attachments. These usually include dividers, a feed wheel and special chopping knives for the long coarse stalks.

6. Alternative delivery chutes for side-loading, hay conditioning or windrowing.

7. Remote hydraulic ram for adjustment of cutting height.

8. Special hitches to simplify coupling and uncoupling of trailers.

9. Balance spring to counter weight imposed upon the rear of the harvester when two-wheeled trailers are attached.

10. Extra large section tyres or twin wheels to prevent sinkage in soft ground.

FIELD PROCEDURE

The flail harvester may be used for various classes of work. The best method of procedure with any one of these can only be determined when precise circumstances are known. Even in the performance of its more usual function, cutting and collecting green crops, the organisation of the work is influenced by many variables such as type of crop to be collected, distance it has to be transported, number of tractors, trailers and men available, and the method of transferring the crop from trailer to silo. Other vital factors such as the desirable degree of laceration or chopping, and the relative merits of direct cutting and collection, or pre-wilting, must also be carefully considered. The management of the whole crop-collection operation is therefore a more involved subject than can be discussed here—indeed, numerous work studies have been conducted on green-crop collection. Clearly, if the harvester is to be kept working continuously, plenty of tractors and trailers must be available to cover the transport and unloading time involved, and close supervision is necessary to avoid bottle-necks.

Actual cutting in the field may be carried out by travelling around the field following the boundary, or by working across the field in bouts after a few cuts have been made around the headland. The former method is the most economical, although if the field is irregular, a few bouts should first be cut adjacent to each side, and the corners radiused. Some operators prefer to eliminate the corner cutting first, by making diagonal runs from corner to corner. Cutting in bouts or lands may be necessitated by the presence of ridges and furrows in the field, when the outfit should travel parallel to the ridges—not across them.

With in-line harvesters, whether working around the field or in bouts, the adjacent cuts should be made in opposite directions to retrieve the crop flattened by one set of the tractor wheels. This means that rotation around the field should alternate, first clockwise, then anti-clockwise. Similarly, adjacent straight bouts should be cut in opposing directions.

If the ground speed has to be reduced because of rough surface conditions, or if gradients cause an involuntary drop in engine speed, a lower gear should be engaged to maintain the required rotor speed. The significance of ground speed/rotor speed ratio has been discussed.

When side-loading into separately drawn trailers, tractor drivers must take care to synchronise their movements—otherwise there may be considerable wastage of the crop. Some loss of material is difficult to avoid in windy weather, although several manufacturers offer extension delivery chutes to reduce this.

When restarting the cut after a hold-up in the work, the outfit should first be reversed so that the rotor can reach its correct operating speed before cutting recommences.

As with other p.t.o.-driven machines, very sharp turns must not be made with the power-drive shaft in operation.

Optimum cutting height will depend upon the nature of the crop and condition of the ground. On stony land, cut a little higher than usual to avoid blunting and damaging the flails, and reduce the risk of stones being thrown towards the tractor driver or being collected with the crop. Even where stones are not present, it is still important to avoid cutting too low and collecting excessive quantities of soil, particularly when cutting green crops from arable land.

When tackling kale, the best crop intake usually results from lowering the rear of the machine on its wheels and lifting the front by the normal adjustment provided. This setting should be accompanied by the lowest rotor speed that provides satisfactory chopping and delivery.

CROP PRODUCTION EQUIPMENT

The collection of a wilted crop needs careful setting to achieve a clean pick-up and the right degree of laceration without blockage. The front apron should be lowered or extended to improve pick-up. However, this adjustment, especially when collecting a crop from arable land in dry weather, can result in a considerable intake of soil, due to the restriction of air intake into the flail housing.

On double-chop machines, the auger which conveys the crop from the flail housing to the secondary chopping chamber can usually be adjusted in relation to its casing. Long and bulky crops usually require more clearance than short, fine materials.

Maintenance

DAILY ATTENTION

The types of bearings used in the construction of forage harvesters vary widely, and each machine should be lubricated precisely as advised by the manufacturer. Some grease nipples and oil-level check points are not always conspicuous, so operators should refer frequently to the instruction book until fully acquainted with the servicing routine.

Correct tyre pressures are vital for consistent cutting height and to eliminate excessive bouncing at high ground speeds. Also, the harvester may have to carry considerable weight when a trailer is attached to it.

On flail harvesters the rotor should be checked for completeness and condition of flails. On chopper-type machines the knives should be examined for condition and positioning.

PERIODIC ATTENTION

The rotor drive belts should be accurately tensioned to prevent slipping and burning, and possibly distorting pulley faces. Frequent checks should be made on the belts of new machines because they tend to stretch. The instruction book should be referred to for the correct deflection as this is always related to the length and section of the belt. The correct tension may be achieved by adjusting a jockey pulley or by moving the gear box. In the latter case, the correct pulley alignment must be maintained. The drive usually employs three or four vee belts, and if one belt breaks or stretches more than the others, the whole set should be replaced to ensure equality of load on each.

The rotor of the flail harvester is dynamically balanced. This ensures that the rotor is free from vibration when centrifugal forces are acting upon it as it rotates at high speed. The slightest condition of imbalance when a rotor is at rest is amplified greatly when it is running at operational speeds. The most likely cause of imbalance is the loss of a flail or the use of unevenly worn flails. Regular checks should be made on the rotor to see that all flails are complete, in good condition and securely attached. Loss or breakage of a flail is usually recognisable by the sudden development of undue vibration throughout the machine. When this happens, a replacement flail should be fitted immediately, and if possible, one with the same amount of wear as those on the rotor. If, however, only new flails are available, and the rotor is of the two- or four-bank type, a pair of flails should be fitted—the second one being diametrically opposite to the replacement one. If the rotor has three banks of flails, three should be fitted, one on each bank (Fig 147a).

As well as its effect on cutting efficiency, the sharpness of the flails has a considerable influence on the power requirement. On very stony land, however, it may be difficult to keep the cutting edges very sharp. The easiest method of sharpening is by grinding, but the cutting edge must not become too hot or the steel will be softened. The bevelled edge should not reach a blue heat, in fact, even the cooler straw colour should be avoided. Some flails are reversible and have two cutting edges.

Methods of attachment of flails to the rotor vary. In some cases, they are bolted or pinned on individually, while in others they are retained by a common pivoting rod to each bank. The latter method simplifies the removal and replacement of a whole bank of flails, but is less convenient for individual replacements. Pivot pins and bushes should be examined for wear when removed. Graphited nylon bushes are favoured by some manufacturers for their reduced rate of wear compared with other types of connection. The shape of some flails necessitates their arrangement in a specific sequence to ensure maximum cutting efficiency across the rotor's width; this will be specified in the maker's instruction book.

The auger drive chain of double-chop machines should be occasionally checked for tension.

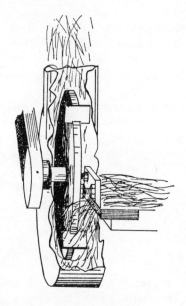

155 Action of flywheel type chopping knife

On double-chop machines, the chopping knives require occasional sharpening, and setting in relation to the stationary shear plate. As with flails, overheating of the knife edges must be avoided when sharpening. Some of these knives are double-edged, to give a double period of service between sharpenings. When setting the knife/shear plate clearance, the manufacturer's instructions should be followed. The actual clearance is usually about $\frac{1}{32}$ in (Fig 155), and it must be uniform along the entire cutting face of each knife. After any adjustments, both the knives and the shear plate must be rigidly secured to prevent either cutting element from becoming displaced when under load.

END-OF-SEASON ATTENTION

Attention to the following items, in so far as they are applicable to a particular machine, can pay dividends in terms of increased life and efficiency.

1. The machine should be thoroughly cleaned, and all bright metal surfaces protected with a rust preventive.

2. Vee belts should be removed and stored away—preferably in a dark place. The belt faces of the vee pulleys should then be coated with a rust preventive.

3. Flails should be checked for condition, sharpened and replaced as necessary and protected from rust.

4. Open bearings should be liberally greased to expel any crop residue which may have entered and accumulated in the bearing housings.

5. Machines should be blocked up to relieve the pneumatic tyres of weight. Also any embedded flints should be extracted from the tyre treads.

6. A further end-of-season recommendation of some manufacturers is to drain the bevel gear box and refill with new oil. Only the specified lubricant should be used.

7. The machine should be checked over generally for worn or damaged parts and for loose frame bolts. The rotor housing skids should be inspected for wear and if necessary reconditioned or replaced.

REPLACEMENT PARTS

The actual spares to be kept on hand will depend upon the type of harvester, but where appropriate the following items should be readily available: flails complete with retaining bolts and bushes, chopping knives, vee belts, shear bolts, slip-clutch elements, chain links. For chopper-type harvesters, pick-up tines should also be available.

Safety precautions

BUCKRAKES

Some of the hazards of using buckrakes for collecting green crops and making silage have been mentioned previously in the chapter, but misuse of this attachment has been and still is the cause of many farm accidents, so no apology is made for the repetition in this summary of warnings. Perhaps the most common cause of accidents in the silage-making operation is that of driving too close to the sides of the clamp, so that the outfit slides to the edge of the clamp and overturns—in many cases trapping the driver. If the ramp of material is approached in a forward direction, steering control is usually lost, and the outfit may slew towards the edge of the clamp. The tractor driver should reverse up the ramp, and avoid the less compacted sides. Because of the labour demands of silage-making, inexperienced operators are sometimes allowed to drive tractors on the clamp to help compaction of the material. This temptation should be resisted both by the employer and by the enthusiastic novice, as handling a tractor on such a precarious 'bottom' requires considerable skill and judgment.

FORAGE HARVESTERS

Several points should be kept in mind when operating forage harvesters.

1. Ensure that all guards are secured in place before starting work.
2. Never attempt to make adjustments in the proximity of moving parts with the machine running.
3. Do not attempt to clear blockages with the machine running.
4. Always handle chopping-unit knives with care, and avoid pushing and pulling on spanners in line with an exposed knife edge.
5. Remember that the blades on the flail-type harvester have a forward-throwing action. When working on very stony land, there

is a tendency for stones to be thrown forward towards the tractor driver, particularly with in-line machines. The same thing may happen when pulverising potato haulms. Some control over this can be obtained by lowering the front apron or by extending the concave.

6. When coupling and uncoupling trailers, avoid standing in vulnerable positions. When driving the outfit for someone else to couple up the harvester or trailer, exercise care in manœuvring into position.

7 MOWING MACHINES

The mower is one of the older items of the farmer's complement of machines and one which finds a place on all types of farm. It is one of the better examples of early farm mechanisation—being introduced during the latter half of the last century. Its reciprocating-knife cutter bar still remains the most widely used cutting mechanism in farm mowers despite the emergence of different cutting systems from time to time. Attention will therefore be focused upon reciprocating-knife mowers in this chapter, though reference will be made to the more recently introduced flail mower and the horizontal rotary mower at the end.

Reciprocating knife mowers

Basically these consist of a cutter bar carried on a fabricated steel frame. The knife drive is powered by the tractor p.t.o. through the medium of a counter-shaft, or by a hydraulic motor.

TYPES AND POSITIONS OF ATTACHMENT

Modern mowers are usually classified as: semi-mounted and

fully mounted—the latter may be either rear-attached or side-attached.

A brief comparative study of these alternatives follows.

SEMI-MOUNTED TYPE

The semi-mounted mower is supported partly by the draw-bar of the tractor and partly by one or two castor wheels at the rear. This close attachment improves manœuvrability and avoids restriction and strain on the power-drive shaft universal joints. The drive is transmitted from the main shaft to the pitman crankshaft by a vee belt or chain.

FULLY MOUNTED TYPE (REAR-ATTACHED)

The fully mounted (rear-attached) machine is totally suspended on the tractor three-point linkage, so that the whole mower is raised when cornering and for transport. This is in contrast to the semi-mounted mower described above which has its cutter bar lift action independent of the main structure of the machine.

FULLY MOUNTED TYPE (SIDE-ATTACHED)

The fully mounted (side-attached) mower is sometimes referred to as the *mid-mounted* type, because of its position midway between the front and rear wheels of the tractor. The obvious advantage of this position is good visibility of the cutting operation. The tractor driver can keep his eye on the cutter bar without hindrance to his steering. Objections to this arrangement were the cumbersome supporting structure and the work involved in attaching and detaching the cutter bar. These have now been eliminated on most models. Operating controls are conveniently placed, and the knife drive is taken from the tractor p.t.o. The abreast position of the mower leaves the rear of the tractor free for attachment of a crimper or tedder so that the two operations can be carried out simultaneously.

Cutter bar assembly and knife drive system

SINGLE-KNIFE CUTTER BAR

The labelled illustration in Fig 156 shows the layout and the various components of a typical single-knife cutter bar assembly. (The term *single-knife* is expressly used to differentiate between

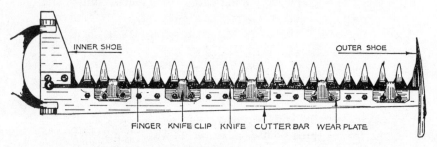

156 Components of cutter bar assembly

this popular type of cutter bar and the more recently introduced double-knife cutter bar to be discussed later.) It is important that the functions of each part of the cutter bar assembly are understood.

CUTTER BAR (FINGER BAR)

The finger bar is the steel plate which forms the basis of the cutter-bar assembly. Most tractor mowers are fitted with 5 ft-wide cutter bars, although some manufacturers are offering 6 ft and 7 ft versions to take more advantage of the capacity of modern tractors. Their sales are restricted however because most swath-treatment machines are designed to handle only 5 ft swaths, though recent designs show a trend towards wider-capacity swath machinery.

Finger bars designed for normal applications are drilled at 3 in intervals to accommodate the fingers, which are bolted in place.

230

FINGERS (OR GUARDS)

Fingers may be of steel or malleable iron. Their functions are to part the standing crop, to support, protect and guide the knife and to act as the stationary cutting faces against which the knife sections make their cut (Fig 157).

KNIFE

Fig 158 shows the construction of the mower *knife*. It consists of a

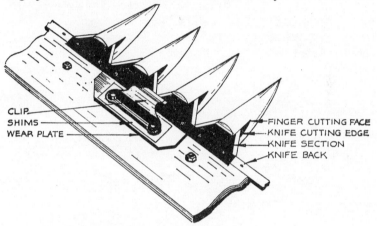

CLIP
SHIMS
WEAR PLATE

FINGER CUTTING FACE
KNIFE CUTTING EDGE
KNIFE SECTION
KNIFE BACK

157 Relationship of knife sections to fingers

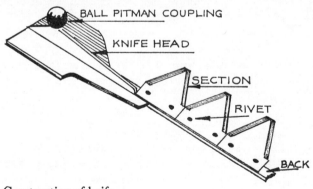

BALL PITMAN COUPLING
KNIFE HEAD
SECTION
RIVET
BACK

158 Construction of knife

231

strip of steel, known as the knife *back*, to which triangular cutting *sections* are riveted. The inner end of the knife is fitted with a pressing known as the *knife head*. The knife head incorporates the ball attachment point for the *pitman* (or connecting rod) which provides the reciprocating motion.

WEAR PLATES

The wear plates are replaceable steel lining plates, designed to protect the finger bar itself from wear by receiving the backward thrust of the knife when in work. They are usually adjustable forward and back to compensate for wear on the knife back and to accommodate the fullness of a new knife when fitted (Fig 157).

KNIFE-RETAINING CLIPS (OR KEEPS)

Knife-retaining clips are spaced at intervals along the cutter bar, and hold the knife sections down close to the ledger surfaces of the fingers on which they slide (Fig 157).

INNER SHOE AND OUTER SHOE

The shoes are usually malleable castings. The *inner shoe* is situated at the inner end of the cutter bar, which attaches to the main structure of the machine, while the *outer shoe* acts both as a divider at the outer extremity of the cutter bar by parting the crop, and as a support for the swath board. Both inner and outer shoes incorporate an adjustable skid on their undersides to determine cutting height (Figs 156 and 159).

SWATH BOARD

The object of the *swath board* is to deflect the outer 18 in or so of the cut swath inwards away from the standing crop, so leaving a margin of stubble to accommodate the offside wheels of the tractor on the subsequent circuit or bout. This margin of cleared land also prevents blockage at the inner end of the knife once the first cut has been made.

The whole cutter bar assembly is attached to the main structure of the mower by two hinge pins which link the inner shoe to the stirrup (Fig 159). This pivoting action allows the cutter bar to

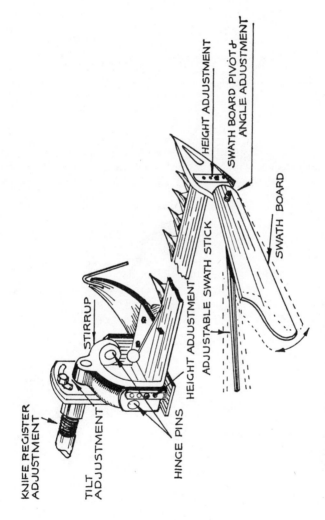

159 Cutter bar hinge point and adjustments

follow ground contours, and facilitates the folding of the cutter bar into a vertical position for transport (Fig 160).

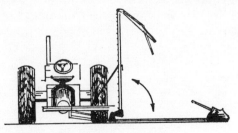

160 Cutter bar in transport and work positions

PITMAN (OR CONNECTING ROD)

The *pitman* converts the rotary motion of the pitman wheel, i.e. the crank-wheel to which it is connected, into reciprocating action at its point of attachment to the knife, Fig 161. Solid and tubular-

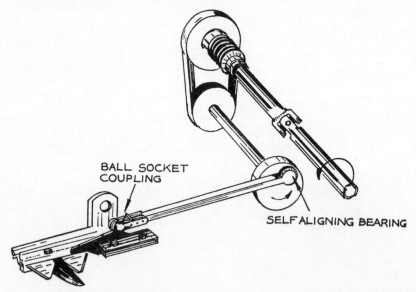

BALL SOCKET
COUPLING

SELFALIGNING BEARING

161 Knife drive system

steel types are available, but wood is favoured by some manufacturers as it stands up well to the peculiar stresses brought to
bear upon it and provides a safety factor in the drive mechanism.
Where wood is used, steel ends are fitted, one embodying a self-
aligning bearing for connection to the crank-pin, and the other an
adjustable ball-socket coupling for attachment to the knife head.
A knife throw of 3 in is adopted on most mowers, and speeds are
usually about 900–1,000 cycles per minute. Relationship of knife
speed to p.t.o. speed is governed by the sizes of the sprockets or
pulleys used in the chain or belt drive between the power-drive
shaft and the pitman crank-wheel. The knife/ground speed ratio
can of course be varied by altering the tractor gear employed.

Correct alignment of the pitman with the knife is important for
mechanical efficiency of the knife-drive mechanism, as also is the
angle of the cutter bar to its direction of travel. These and other
important considerations are referred to in more detail under
Maintenance.

DOUBLE-KNIFE CUTTER BAR

Fig 162 illustrates a double-knife cutter bar. In this case, two
knives are fitted, one above the other. Their actuating pitmans are
driven by a two-throw crank whose crank-pins are diametrically

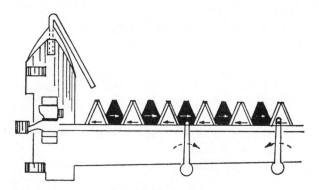

162 Double-knife cutter bar assembly: no static fingers, both sets
of knife sections reciprocating

opposed. This causes the knives to reciprocate in opposing directions (Fig 163). In this case, therefore, the cutting is achieved by two sets of moving cutting elements, instead of one moving and one stationary set of cutting edges as in the single-knife cutter bar. The extra cutting efficiency permits a higher ground speed,

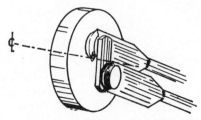

163 Two-throw crank of a double-knife mower

and in suitable conditions is capable of a closer cut than the single-knife cutter bar. This double-knife assembly has fewer wearing parts on the cutter bar itself, but the knife-actuating mechanism is more complicated. The opposing movements of the knives reduce cutter bar vibration.

A recent development, being introduced on some mowers, is the hydraulic drive to the knife-actuating mechanism.

MECHANICAL PROTECTIVE DEVICES

A slip clutch is usually incorporated in the knife-drive system to protect it from overload damage. This may be omitted if the machine employs a vee belt drive, in which case slippage of the belt(s) provides a measure of protection. If steel pitmans are used, they usually incorporate shear bolts, to protect the drive system from serious shock load if an obstruction is encountered.

All modern mowing machines incorporate a cutter bar release device, sometimes known as the *break-back*. This permits the cutter bar to swing back if it strikes an obstruction, so avoiding damage from the impact (Fig 164).

236

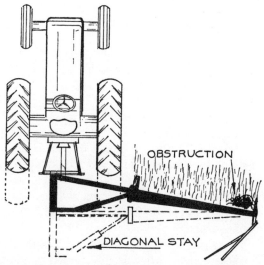

164 Typical cutter bar break-back device: diagonal stay telescopes and relieves impact

Attachment to tractor

SEMI-MOUNTED AND FULLY MOUNTED TYPES

The attachment procedure for both semi-mounted and fully mounted mowers is too individual and specific for general comments to be of much value. The attachment of some semi-mounted machines requires vertical hinging provision to accommodate undulating land. Others attach directly to the three-point linkage, as with other rear-attached fully mounted machines. Adequate instructions are issued by the makers, giving precise details for each make and model of mower and tractor. Such recommendations sometimes include the setting of the tractor wheels to a specific track width.

Operational adjustments

Reference has been made to some operational adjustments, but all possible adjustments should be considered in more detail.

237

CUTTER BAR LIFT CONTROL

The lifting of the cutter bar above the swaths of cut material at the turns is now generally done through the tractor hydraulics. Fully mounted mowers are usually lifted as a whole. On some semi-mounted mowers the cutter bar is raised independently of the main structure of the machine, either manually or mechanically, and counterbalance springs relieve some of the cutter bar weight. This reduces the effort required to raise it, and helps it to float on the ground surface when in work, so eliminating drag and excessive wear on the shoe skids. Balance springs are also fitted to give this flotation on mowers without independent cutter bar lift.

Adjustment is usually provided on the cutter bar lift linkage, to counteract the slight tendency of the outer end of the cutter bar to sag. Specific measurements are normally quoted by manufacturers.

CUTTING-HEIGHT ADJUSTMENT

The cutting height is set by adjusting the skids on the underside of the inner and outer shoes at the ends of the cutter bar (Fig 159). When this adjustment is modified, the cutter bar should have equal clearance from the ground at either end to give an even length of stubble.

TILT OF CUTTER BAR

By tilting the cutter bar, the fingers can be given a slight downward or upward inclination in relation to the ground surface. In normal cutting conditions, it is usual to set the bar so that the knife sections are level. Where stones are present, as they may be when cutting a temporary ley on stony land, they should be angled slightly upwards to encourage the fingers to ride over the stones. In such circumstances the balance spring should also be adjusted to carry more weight. For very close cutting, on the other hand, it is sometimes advantageous to tilt the fingers slightly downwards, although in soft ground conditions, or in badly

9. Off-set flail harvester picking up windrowed crop

10. Chopper-blower harvester picking up windrowed crop

11. Tedder using over-top action

12. Rake-wheel type machine swath-turning

ridged fields, this may result in the points of the fingers digging in. The adjustment is usually by hand lever, though on some mowers a spanner is necessary for making this setting. The cutter bar tilt on mowers designed for attachment to the tractor three-point linkage can also be influenced by the length of the top link.

KNIFE SPEED

Many mowing machines have only one knife speed, and this is satisfactory over a wide range of crops and conditions. Some makers, however, make two or more speeds available by fitting stepped vee pulleys. Knife speed should only be varied in accordance with the maker's recommendations, as excessive speeds can damage the knife-drive mechanism. The ability to vary the knife speed/ground speed ratio on power-driven mowers by tractor gear selection has been previously mentioned.

SETTING OF SWATH BOARD AND STICK

The only possible adjustment on the swath board is its angle in relation to the cutter bar. This is by varying the tension on the spring at its attachment point (Fig 159). Adjustment of this can sometimes help to speed up the drying of heavy crops, or reduce the bleaching of light crops, by increasing or decreasing the angle of the board respectively in relation to the cutter bar to alter the width of swath it leaves. The swath stick should be set to deflect the crop and prevent it overshooting the board.

Field procedure

TRANSPORTING THE MOWER

When preparing the mower for transport, ensure that the p.t.o. drive is disengaged, before folding the cutter bar into its vertical position. Also, before driving on the highway, cover the cutter bar fingers completely, because of the hazard they otherwise present to other road users.

I

CONDITION OF CROP

The timing of cutting operations depends upon a number of factors, such as the nature of the crop, its stage of maturity, the intended method of conservation and weather prospects. Generally speaking, the earlier a hay crop is cut, the higher is its protein content and general nutritional value, although quantity of yield has to be sacrificed for quality, and drying of the crop may be slower. The cutting efficiency of the mower is not normally upset by damp conditions, although the advisability of proceeding with work in unsettled weather is another matter.

The cut should be reasonably close, but a stubble of about 2 in is preferable to accelerate aftermath recovery. Stony conditions will require a higher cutter bar setting.

OPENING OUT THE FIELD

Work usually starts at the perimeter of the field, with the outfit following the boundary hedges or fences. The fact that the cutter bar is offset—projecting beyond the side of the tractor—necessitates a *back-swath*, i.e. a cut made in the opposite direction to the normal direction of cutting. With the conventional right-hand cutter bar normal cutting is clockwise and the back-swath will be anti-clockwise. This may be cut either on the initial run round the field with the cutter bar immediately adjacent to the hedge, or after work has commenced in the normal clockwise direction. In either event previously cut grass will probably block the inner end of the cutter bar, although this can be obviated where a second swath board can be fitted on the inner end of the cutter bar for the initial cut.

Since the introduction of the flail-type forage harvester, many farmers have been opening out their mowing grass by making a few rounds with this machine. So long as good use can be made of the collected material, this has the advantage of both simplifying the opening out of the field for mowing, and eliminating the problem of drying the crop lying close to hedges.

CORNERING

Once the field has been opened out, mowing can proceed, with the outfit either 'looping' the corners (turning left-handed through 270°), or turning right-handed sharply at each corner of the standing crop. When cutting conditions permit a relatively high ground speed the outfit should be slowed down at the corners as it passes over the cut swaths, to avoid excessive cutter-bar whip.

Reference has been made to the importance of knife speed being kept within the range recommended by the manufacturer. For efficient cutting without undue strain on the machine, ground speed must be regulated to suit the density and toughness of the crop. Up to 6 miles an hour is permissible with most mowers in good cutting conditions, although higher speeds are sometimes possible—especially with those machines having double-knife cutter bars. At all times tractor gear selection is important to ensure a suitable knife speed/ground speed ratio.

OPERATIONAL CHECKS

Inefficient cutting or knife blockage may result from dull cutting edges on the cutter bar components, or from incorrect alignment and installation of the knife. Repeated 'breaking back' of the cutter bar safety-release device indicates undue pressure on the cutter bar. This may be caused by inadequate knife speed, excessive ground speed or poor cutting performance (this assumes of course that the safety release itself is properly adjusted). Factors influencing cutting efficiency are detailed under *Maintenance*.

Uneven cutting height across the width of the cutter bar is due either to unequal setting of the skids at the shoes, or to restriction on the cutter bar lift linkage.

Maintenance

DAILY ATTENTION

The lubrication requirements of mowers varies with the form of power transmission used and the types of bearings fitted. Lubrication of the power-drive shaft itself, however, and of the pitman shaft, is usually a daily requirement. Lubrication of the cutter bar components, on the other hand, is seldom recommended—never, in fact, for work in dusty conditions, although it might be permissible when cutting meadow grass from a thick sward, where there is less likelihood of abrasive soil being collected on the lubricated surfaces.

The tension of vee belts should be checked frequently, especially when cutting heavy crops.

The period of satisfactory service obtained from a knife before it requires resharpening depends on the density and coarseness of the crop. Suitable methods of sharpening are described later. It is usual to carry one or two spare knives for quick replacement in the field, as and when necessary.

Tyre pressures should always be maintained as specified.

END-OF-SEASON ATTENTION

After a season's work, a relatively new machine will normally require little more than cleaning, a general check over, sharpening of the cutting edges and protection from rust. On a machine which has seen several seasons' work, the cutter bar may require a general overhaul and checks on its alignment. This is probably best undertaken by an agricultural mechanic, although many experienced tractor drivers are capable of carrying out this work on the farm where suitable facilities are available. With this in mind, an outline is given of the procedure.

Overhauling the cutter bar

The various operations and methods of procedure entailed in an overhaul of the cutter bar are as follows:

Reconditioning the knife

When knife sections are sharpened, the original angle and bevel of the cutting edges must be maintained or efficiency may be reduced.

A common fault is to increase the angle by filing heavily near the apex of the section (Fig 165a). Also, if the bevel is too steep (Fig 165c), the cutting load will be excessive, if too shallow (Fig 165b), the edge will soon become blunt or burred because of insufficient *backing*. *Smooth* files used with a semi draw-filing action, or hand carborundum stones, are generally the most effective for sharpening, although special grinders are available incorporating a jig for semi-automatic operation. These are quick and efficient, but care must be taken not to overheat the sections and soften them. Un-

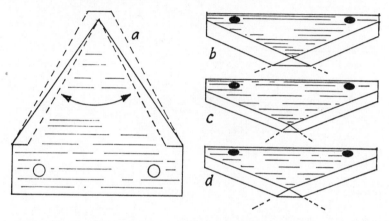

165 Angle and bevel of knife sections: *a* Excessive filing at apex, *b* Bevel too flat, *c* Bevel too steep, *d* Bevel correct

necessary grinding or filing should be avoided although it may be necessary to first true up chipped cutting edges. When repeated sharpening has seriously reduced the width of the sections, they should be replaced, or cutting action may be lost at the extremities of the knife's stroke.

Overlapping type sections are best removed as shown in Fig 166; the section of the knife to be removed is placed between the slightly opened jaws of a vice, while the back rests on the top of the rear jaw. One sharp blow with a hammer on the edge of the section, immediately over each rivet, is usually sufficient to shear the rivets. To remove the less common flush-type sections, a chisel should be used to cut off the rivets, with the knife-back held firmly in the vice. A special hand-powered machine for removing and replacing sections may be justified on a farm where much mowing and combine harvesting is done. It simplifies the removal of knife sections, and re-rivets new ones very rapidly.

Knife sections vary slightly in some of their dimensions, so it is important when purchasing spares to check that they are of the

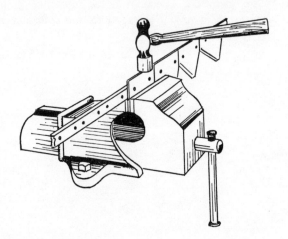

166 Removal of knife sections

correct pattern. Variations occur in the diameter and the spacing of the rivet holes. The rivet length is also important if the stem is to be properly expanded to ensure a sound job with a neat finish. A projection of $\frac{1}{4}$ in of rivet stem is satisfactory, especially if, as is preferable, a proper rivet *set* is to be used instead of direct hammering.

Serrated section knives are available, though not commonly used for grass cutting, and special fingers are obtainable with serrated side faces on their ledger plates.

Reconditioning the fingers

During the cutting season, many mower knives are sharpened

enthusiastically by operators, whilst the necessity for a good cutting edge on the finger is often overlooked. This is tantamount to sharpening only one blade on a pair of scissors and expecting good results. The fingers do not need to be sharpened as frequently as the knife sections, but they must have a keen edge at the shearing face (Fig 167). On steel fingers, this face is an integral part of the forging, but on malleable fingers, a hardened steel *ledger plate* is riveted in place, as malleable iron has poor wearing qualities. Grinding is the most effective way of sharpening the fingers; for

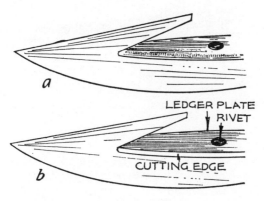

167 Condition of finger cutting faces: *a* Blunt, *b* Sharp

accurate work, they must be removed from the finger bar. The angle of bevel should be a little less than a right angle with the top face—about 75° to 80°. After repeated grindings, fingers without ledger plates will need replacing, but malleable fingers can be reconditioned by fitting new plates.

The fingers must be aligned so that the knife runs smoothly, without snaking as it reciprocates. With all fingers bolted securely on to the finger bar, a superficial check may be made by sighting along the horizontal faces of the fingers from the end of the bar. Any misalignment is shown up more clearly if a thin piece of twine is secured at the inner shoe and drawn taut as indicated in Fig 168. This alignment of the *horizontal faces* of the fingers is more critical than true alignment of their points. Correction where necessary can be made by packing with shims, or by carefully levering or tapping the finger back into alignment.

168 Using line to check finger alignment

Installing the knife

Even with well sharpened knives and fingers, good cutting is not possible if the knife is incorrectly set in relation to the fingers. This setting involves precise adjustment of the knife-retaining clips and wear plates. Fig 169 illustrates correct and incorrect conditions. Correct excessive clearance as follows:

First, fit the knife into the cutter bar, and check the vertical clearance between the knife sections when held down on the finger faces, and the underside of the knife clip. If shims have been provided under the clips by the manufacturer, remove one or two until the correct clearance is obtained. Otherwise, remove the knife and tap the clips carefully down, using a spare knife section to make frequent checks. Recommendations on the clearance vary, but try to obtain a sliding clearance without undue pressure or binding of the knife.

When all the clips have been checked, slacken their securing bolts so that the wear plates—positioned beneath the clips—may be adjusted. To do this, push the knife forward to take up any play, and move the wear plates up and secure. The knife should then be checked to ensure that it slides reasonably freely in the cutter bar, i.e. it should be possible, with a reasonable amount of effort, to move the knife by hand through its operational strokes.

Replace any retaining clips or wear plates that are unduly worn.

In view of the importance of correct knife installation, do not alternate old knives having badly worn backs with new knives.

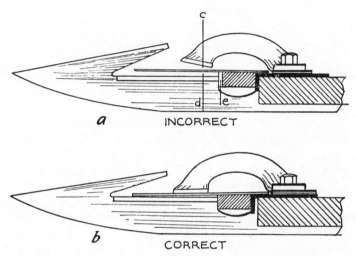

169 Knife clip and wear plate adjustment. Note excessive clearance at *c*, *d* and *e*

It is good policy to buy knives in sets of three, and to interchange them regularly to maintain even wear and good interchangeability.

Finally, check the knife head for undue play in all possible directions, i.e. *vertical*, *forward* and *back* (Fig 161). A little correction is sometimes possible by moving the rear knife guide forward on the inner shoe, and lowering the holding-down plate positioned above the knife-head front guide-plate. Excessive play at this point causes undue strain at the inner end of the knife back —a fault which usually gives rise to pronounced rattling, and in extreme cases, to knife breakage.

247

Checking the pitman

One end of the pitman has a bearing which may be either ball, roller or bronze bush and the other carries the knife-coupling socket. Both should be checked for wear and replaced if necessary. Premature wear on the coupling socket is often caused by over-tightening, while on the other hand, a slack connection results in the socket depressions becoming elongated.

Checking cutter bar lead and pitman alignment

Some instruction books supplied with mowing machines refer briefly to cutter bar lead and pitman alignment checks, without explaining their meaning and purpose.

The term *cutter bar lead* refers to the angle of the cutter bar in relation to the mower's direction of travel. For operational efficiency, when the machine is in work this should ideally be at right angles, but clearly, as the cutter bar is offset, projecting to the right of the point of the mower's attachment to the tractor, the resistance of the crop tends to force the cutter bar back, particularly in heavy cutting conditions. This has two effects: first, it tends to slew the whole machine round so that the cutter bar is no longer at right angles to the outfit's direction of travel. Second, it forces the cutter bar back from its hinged point of attachment to the main structure of the machine (Fig 159). So, there are two forms of misalignment which can occur, the first of which has a slight effect on cutting efficiency, while the second has a more serious effect on the mower's mechanical efficiency; indeed, the significance of this demands more detailed explanation.

If the outer end of the cutter bar lags behind the inner end from its hinge pins, at each outward stroke of the pitman, the head of the knife (inner end) tends to kick forward and unduly strain the knife back. This can be so severe as to eventually break the knife.

If, on the other hand, the cutter bar has *lead*, i.e. its outer end set slightly forward of its inner end, any tendency of the knife head to kick on its outward stroke will be rearwards, but in this event the thrust will be received without harm by the knife-head guide. It is therefore desirable for the cutter bar to have lead.

However, when the cutter bar has lead, the knife must inevitably be slightly out of alignment with the pitman, but this minor discrepancy is accommodated by the self-aligning nature of the pitman inner-end bearing and ball-knife coupling.

When wear develops at the hinge pins, cutter-bar lead also ensures that the maximum pitman-to-knife misalignment occurs when the machine is *not* under load, and conversely, the minimum misalignment exists when the cutter bar is subjected to maximum crop pressure.

Unfortunately, most instruction books oversimplify the interpretation of cutter bar lead, and specify a definite measurement, implying that the cutter bar moves back under pressure to its right-angled position. In fact, cutter-bar lead measurement can only be arbitrary, since the resistance of one crop can be twice that of another. Also, ground speed and balance-spring adjustment can further influence the rearward pressure and drag imposed upon the cutter bar. These facts are reflected in the inconsistency of lead measurement recommendations by different manufacturers—ranging from $1\frac{1}{4}$ to $3\frac{1}{2}$ in for a 5 ft cutter bar!

Forms of lead adjustment

On some mowers, provision is made only for angling the machine as a whole, or for positioning the stirrup and its supports together with the pitman and cutter bar. On other machines, the cutter bar can be angled relative to the pitman, independently of the stirrup and its supporting bar or frame, by an eccentric-spigoted hinge pin (Fig 170). For the reasons outlined above, this is the more effective of the two settings. Sometimes, both forms of adjustment are possible. Where this is the case, the following procedure may be adopted to set the cutter bar correctly.

Checking and setting pitman alignment

The pitman should be at right angles to its actuating crank-shaft. This can be checked by placing a straight-edge, approximately the

249

length of the pitman, across the front face of the pitman crank-wheel (Fig 171). If the pitman is not parallel with this, correct the alignment by lengthening or shortening the stirrup-supporting stay(s) (Fig 170).

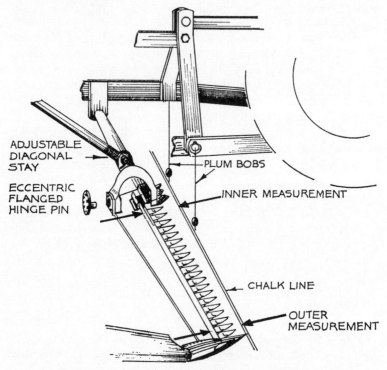

ADJUSTABLE DIAGONAL STAY

ECCENTRIC FLANGED HINGE PIN

PLUM BOBS

INNER MEASUREMENT

CHALK LINE

OUTER MEASUREMENT

170 Checking lead of cutter bar in relation to tractor

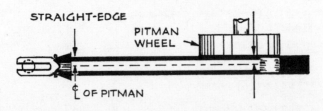

STRAIGHT-EDGE

PITMAN WHEEL

℄ OF PITMAN

171 Checking pitman alignment by means of a straight-edge

Checking and setting cutter bar lead

When set as above, the pitman may be used as a datum to check the alignment of the cutter bar. A line secured at the crank-wheel end of the pitman should be pulled taut from a point beyond the outer shoe of the cutter bar, and lined up central or parallel with the length of the pitman (Fig 172). By measuring between the line and the rear edge of the knife sections at each end of the cutter bar, the existing lead (or lag) of the cutter bar relative to the pitman may be ascertained.

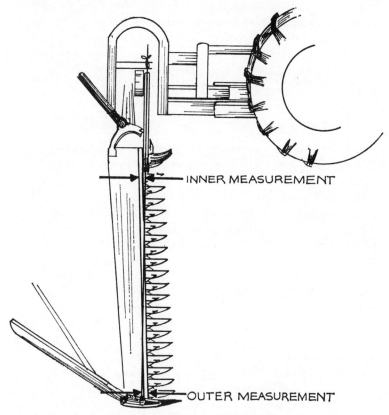

INNER MEASUREMENT

OUTER MEASUREMENT

172 Checking lead of cutter bar in relation to pitman

In spite of the arbitrary nature of lead specifications, it is as well to comply with the recommendations given by the makers. The usual form of adjustment for independent cutter bar lead has been mentioned—namely the eccentric hinge pin, which rotates to alter the angle of the cutter bar (Fig 170). If the cutter bar lags beyond the scope of the adjustment, excessively worn hinge pins and bushes, or distortion of the stirrup or its supporting stays, may be responsible.

Overall lead of the entire machine, or of the cutter bar, stirrup and pitman as an assembly, relative to the tractor, may be checked by projecting a chalk line established on a concrete floor by dropping a plumb-bob from the rear attachment points of the two lower links (Fig 170). Once the correct setting has been obtained, the distance between the point of the outer shoe (with the cutter bar lowered to the ground) and the centre of the offside rear tractor wheel can be measured and noted, to simplify re-checking whenever a discrepancy is suspected.

Knife register

Reference has been made to the fact that most mower knives have a 3 in throw. It will also be remembered that the fingers are spaced at 3 in intervals. So the completeness of the cut depends on the starting and finishing points of the knife's movement. On most cutter bars, the knife movement is arranged so that the sections register centrally with the fingers at the extremities of each stroke, that is the centres of the sections line up with the centres of the fingers (Fig 173). This arrangement is by far the most common, but it does result in a small amount of ineffective knife movement near the end of each stroke. An alternative design, which eliminates this dead knife movement by means of a slightly shorter working stroke, has been used on some machines, but has not been widely adopted. The knife throw in this case terminates just short of the centre of adjacent fingers. Because of this it is advisable to determine the specified knife throw for any particular machine before attempting to re-set its knife register.

It is unlikely that the knife register will become disturbed in the normal course of work. However, it must be checked whenever a new pitman or knife is fitted, as a slight discrepancy in the length of the pitman, or a slight variation in the location of the knife-head ball, can upset the register. At the present time, few machines are fitted with pitmans of adjustable length, so that adjustment of the knife register necessitates moving the whole cutter bar

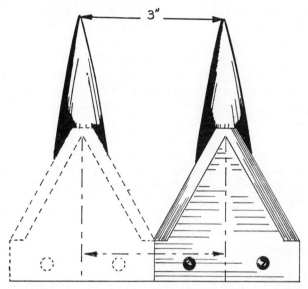

173 Knife register where knife has three-inch throw. Sections should centralise on fingers at the extremity of each stroke

assembly relative to the knife. Various ways of doing this are found on different machines, one of which is illustrated in Fig 159. In any event, the adjustment is usually simple to carry out with the guidance of the instruction book. The adjustment should be made with the cutter bar lying flat on a level floor in its normal working position.

Pitman-drive mechanism

The chain or belt drive should be periodically inspected and its tension checked. Wherever a protective slip-clutch is incorporated in the knife-drive system, this should be checked to ensure that it will function if the mechanism is overloaded.

Cutter bar lift linkage, balance spring and break-back release

Certain adjustments are possible on the cutter bar lift linkage and the methods and measurements are given by the manufacturer.

The balance spring should be set to carry as much of the cutter bar's weight as possible without allowing it to bounce. A *light* cutter bar is desirable when working on stony land although a *heavier* bar, where permissible, generally gives a more regular cutting height, especially where high operational ground speeds are used.

The cutter bar break-back device should be kept in free working condition throughout the season. Tension on the spring-release mechanism is adjustable and this should be set *lightly* at first, and then modified under working conditions. It should be protected from rust throughout its period of storage.

REPLACEMENT PARTS

The mowing machine initiates the haymaking campaign which is so vital on most farms, and so its mechanical reliability is of paramount importance. Even where the machine has been duly checked over, the following parts should be on hand at the start of a season's work: at least two knives, two or three fingers, a supply of new sections with rivets of the correct lengths for both normal section and knife-head riveting, pitman-drive belt(s) or chain (as applicable), and one complete pitman with spare shear bolts if these are incorporated.

Safety precautions

The danger points on the mowing machine are more obvious than those of many other pieces of farm equipment. Nevertheless the following points are worth keeping in mind.

1. Always exercise care when handling and sharpening knives.

2. Do not attempt to clear a choked cutter bar with the knife in operation.

3. Disengage the p.t.o. drive and lower the cutter bar to the ground whenever the tractor engine is stopped.

4. When folding or unfolding the cutter bar for transport or work, avoid placing the hand or fingers between the fingers of the cutter bar.

5. Before travelling on the road, fix the cutter bar securely in the vertical position and cover the fingers with a guard. Any spare knives should also be carried in a guard and secured to the tractor or mower.

Flail mowers

The flail mower came into being fairly recently following the success of the flail harvester—indeed its structure and working principles are almost indentical with the flail harvester (see Fig 174). One important difference, however, is the absence of intensive laceration fitments on the flail mower.

Already it has been proved to be a great success both in mechanical reliability and in its effect on the crop.

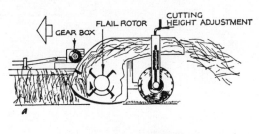

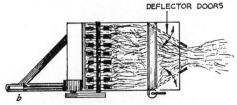

174 Flail mower: *a* Side elevation, *b* Plan view

It employs a lower peripheral rotor speed range than the forage harvester, as the aim is not to lacerate the material severely but merely to bruise and fracture it sufficiently to accelerate its subsequent drying in the swath. The reasoning behind this is fully explained in Chapter 8 under *Conditioners*.

Both 5 ft and 6 ft cut versions are available and it can cut to within 2 in of the ground. On some machines the cutting height is set hydraulically. Deflectors are available to regulate the position of the swath. In normal working conditions, stoppages are exceedingly rare—indeed the flail mower is tolerant of the most difficult conditions. In some cases the wheels can be positioned either alongside the rotor for true following of contours or behind the rotor to avoid running on the uncut crop.

Maintenance requirements are similar to those of the flail forage harvester.

HORIZONTAL ROTARY MOWER

This is less commonly used in agriculture. The cutting element is a heavy horizontally rotating disc, having cutting blades attached to its periphery. It can cope with long grass and with uneven ground conditions. Its power requirement is comparatively low and the cutting element is less prone to serious damage from obstructions—the small cutting blades are easily and quickly replaced.

8 HAYMAKING MACHINERY

The British climate makes the production of good quality hay one of the most difficult operations on the farm, especially if the crop is harvested by the traditional method of drying in the field. However, in spite of the weather, and notwithstanding the impact of the extensive and ever-increasing interest in the conservation of green crops by ensiling, hay still remains the mainstay of stock fodder. Hay is a vital crop, and farmers are not lacking in readiness to modify their techniques to improve its quality, so much so, that new thinking is revolutionising the equipment used in haymaking operations.

The general procedure involved in the drying of hay is fairly common knowledge, but there are many less obvious factors which can make or mar a potentially good crop. Some of these are, to some extent at least, within the farmer's control, and it is worth analysing the various necessary processes, and the problems they present.

The traditional reciprocating-knife mowing machine leaves the cut crop, which may have a moisture content of up to 80%, lying on the ground in flat swaths. Before the crop can be collected for stacking without risk of deterioration, this moisture content must be reduced to about 20%. Even before hay can be baled—

the more usual practice at the present time—the moisture content must be reduced to about 30%. So, to condition a crop for collection, a very large quantity of water has to be evaporated, and this, quite often, in a catchy season. Moreover, the optimum time of cutting often does not coincide with suitable settled weather, and rain on the crop during its various phases of curing can considerably reduce its nutritional value by *leaching*, i.e. washing out of valuable nutrients.

A further problem is mechanical damage. Much of the protein of the crop is contained in the leaves of the herbage, and repeated mechanical handling tends to strip and shatter these from the stems, particularly at the later stages of drying. It is obvious, therefore, that careful selection and timely and efficient operation of swath-treatment machinery is very important.

These problems of urgency and delicacy in the treatment of hay have led to the introduction of the wide range of contrivances at present available. The fundamental principles of these machines vary so widely that selection is not easy. However, we will first consider the various processes they are designed to perform.

Swath treatments

The standard techniques employed in haymaking are generally known as tedding, turning, spreading, gathering, side-raking, and conditioning. These are not arranged in order of sequence, neither are all these treatments likely to be applied to any one crop.

TEDDING

The word *tedding* has the same basis as the Welsh word *teddu*, which means *to spread out*. However, so far as hay is concerned, the term implies more than just spreading the swath. The precise treatment depends upon the mechanical action of the machine used. The general objective with the tedder is to lift the swath left by the mower, throw it to a greater or lesser extent, and leave it in an open 'fluffy' condition, so that air can freely circulate through it (Fig 175). Given suitable weather, tedding usually follows closely behind the mower.

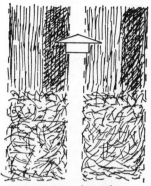

175 Tedding—vigorous aerating of swaths

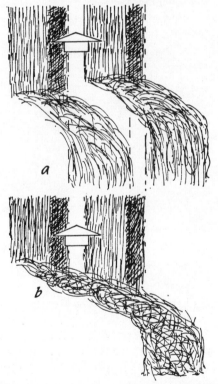

176 *a* Timing—inversion and movement of swaths or windrows.
b Side-raking—side-sweeping effect placing two windrows into
one

TURNING

Turning is self explanatory. When one side of the swath has been exposed to the sun and drying wind, it must be inverted to expose its damp underside. The action of most machines is such that the operation also moves the swaths sideways on to drier ground (Fig 176a).

SPREADING

In *spreading* the crop is scattered to give faster and more even drying. Spreading is sometimes used, as an alternative to tedding, direct on the mown swaths or on swaths which have already been moved, but which have been soaked by rain. The spreading action is less severe on the crop than tedding (Fig 177).

177 Spreading—less severe scattering of swaths or windrows

GATHERING

In the *gathering* operation, two swaths are gathered together into one. This is also true of side-raking, but the term *gathering* usually implies that the machine is capable of centre-gathering, i.e. collecting a swath from each side of its working width and discharging it at the centre (Fig 178). As the crop travels a shorter distance laterally, 'roping' is not so likely to occur with centre-gathering as it is with side-raking.

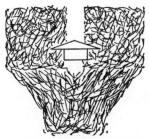

178 Gathering—the action re-windrows scattered or tedded crop

SIDE-RAKING

Side-raking moves the crop laterally—usually handling two swaths at a time—and discharges the resultant windrow to one side of the machine (Fig 176b). The windrows so produced are usually neat and tidy, although there is sometimes a tendency for long material to be tightly rolled.

CONDITIONING

Although in a sense all haymaking machinery is used to condition the crop, in a comparative study of equipment the term *conditioning* implies physical processing of the crop as distinct from merely moving or aerating it. Slightly differing processes are used and these generally fall under the headings—crimping, crushing and lacerating.

Crimping

Crimping is comparatively new so far as British haymaking techniques are concerned, but it has made considerable impact since its introduction. The swath is passed between two grooved or fluted rollers to crimp or kink the stems of the crop at intervals of about 2 or 3 in (Fig 179a), to speed up their drying rate. This helps to overcome the problem of unequal drying of leaf and stem which results in the loss of the protein-rich leaf before the stems are dry. For best effect, crimping should be carried out

261

immediately behind the mower, or at least within half an hour of the actual cutting. As well as promoting more uniform drying in good haymaking weather, the crimping process also accelerates overall drying, a considerable asset in less sunny conditions. Particular crop or weather conditions may necessitate repetition of the treatment, although the roller pressure must be carefully adjusted to avoid undue severity.

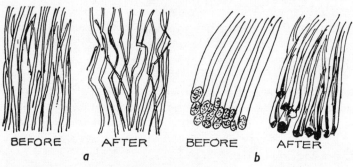

BEFORE   AFTER   BEFORE   AFTER

a   *b*

179 Effect of *a* Crimping, *b* Crushing

Crushing

Crushing is similar to crimping, in that the swath is passed between rollers to accelerate the drying of the stemmy portion of the crop. In this case, however, the rollers are usually almost entirely plain, and so crush or split the stems throughout their entire length (Fig 179b). Some users prefer this for thick-stemmed leguminous crops, both for the advantage of quicker drying and for improved digestibility of the resultant hay.

Lacerating

In this case the crop is fractured and bruised by flails. As an alternative to investing in a special hay conditioning machine, some farmers use the flail harvester, with modifications to reduce the lacerating effect and the crop discharge level (Fig 154). Reference has been made to this in Chapter 6.

The latest innovation is the flail mower described in Chapter 7, which cuts and conditions the crop in one operation. It can also be used to repeat the conditioning treatment after the initial cutting.

Construction and operation of swath-treatment machines

There is a multiplicity of machine types available, and classification is made difficult by their varying degrees of versatility. Most machines, whether labelled by their manufacturers as tedders, swath turners or side-rakes, are either dual or multi-purpose units. For instance, some tedders can also spread and turn swaths, and some swath turners can ted and side-rake. For this reason, classification might be clearer if determined by the type of mechanism employed in the various machines.

REEL TYPE

Fig 180a illustrates a reel-type machine. Several versions are available, both p.t.o. and ground-driven, and these are distinguished by their action on the swath, i.e. over-top, back-kick, combined over-top and back-kick, angle tedding or side-raking. Some machines can perform all these operations. Another distinguishing feature is the working width; some machines are designed for single-swath treatment and others for double-swath. Machines of this type are also being produced to satisfy the increasing interest in wider swaths than the hitherto almost universal 5 ft width.

Over-top action

Over-top action implies that the reel rotates in the opposite direction to the land wheels, so lifting the swath, carrying it over the top of the reel periphery and throwing it upwards and rearwards, leaving a high open windrow (Fig 180b). This action provides maximum movement and aeration of the crop and is suitable for the early stages of drying. Selection of ground speed and reel speed gives some control over the severity of the treatment, and special care must be taken when tedding a crop in its later stages

263

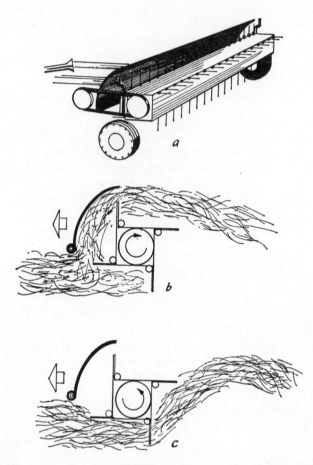

180 Reel type swath treatment machine: *a* General structure.
Working with *b* Over-top action, *c* Back-kick action

of drying. On some ground-driven machines, reel speed can be
varied, and on power-driven machines, the treatment can be
regulated by careful selection of the tractor gear. In some cases,
intensity of tedding can be regulated by an adjustable feathering
action on the tines. This also discourages the tendency of long
material to wrap (Fig 181). Strong treatment is generally necessary

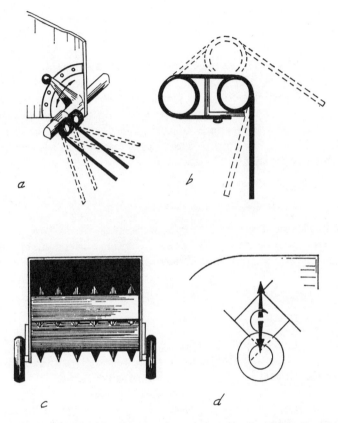

181 Tedder tines—*a* Tine angle adjustment for feathering or flicking action, *b* Double-coil spring tine—relieves strain in two directions, *c* Enclosed square drum tedder with steel teeth—produces a valuable air draught to 'fluff-up' windrow, *d* Floating square drum rotor—to follow land contours

for newly cut grass, and gentler handling for material which is partially dry. Plate 11 shows a reel type tedder working with over-top action.

Back-kick action

For *back-kick action*, the reel must rotate in the same direction as the land wheels. The backward movement of the tines kicks the swath upwards and rearwards from the lower region of the reel periphery (Fig 180c). This treatment is less vigorous than the over-top action, and is therefore more suitable for late-stage tedding, although not as effective for early stage treatment as the over-top machine, especially in a heavy crop.

Alternative over-top/back-kick action

On a machine designed for both over-top and back-kick action, the reel can be made to rotate in either direction, so that either action can be selected according to the condition of the crop.

Angle-tedding action

Many reel type tedders, whether of the over-top or back-kick type, can be used for *angle tedding*, i.e. tedding with the reel set obliquely in relation to the direction of travel. This delivers the swath offset from its original position, placing it on fresh, dry ground. It also spreads the swath laterally to a greater or lesser extent according to its angle of set (Fig 182a). This is useful for moving swaths from under hedges and trees further out into the field to aid drying.

Side-raking action

A side-raking action is possible with some reel-type tedders. As with angle tedding, the machine is angled, but for side-raking, the reel must rotate in the reverse (over-top) direction (Fig 182b).

The attitude of the machine is usually set by altering the angle of the land wheels relative to the frame of the tedder.

In contrast to the tedders so far illustrated, some machines have the reel more enclosed in a sheet-metal housing. Deflector doors are fitted at the rear and these can be adjusted to regulate the discharge of the hay.

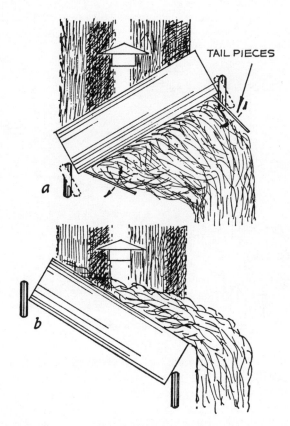

182 Tedder used for *a* Angle tedding, *b* Side raking (reel rotation reversed)

RAKE-WHEEL TYPE MACHINES

Fig 183 illustrates a rake-wheel (or finger-wheel) machine. As will be seen, it comprises a frame supporting a series of rake wheels arranged obliquely to the outfit's direction of travel. The rake-wheels are mounted to rotate freely on their central spindles, and are turned by ground contact. The spring-steel tines projecting beyond the rims of the wheels give some measure of flexibility as the machine passes over minor land undulations. On some machines

183 Rake-wheel type machine (side-raking)

the rake-wheels can rise and fall independently. The rake-wheel bearing assemblies can be swivelled to alter the angle of their approach to the crop, so enabling different operations to be performed, such as turning, tedding, side-raking and gathering—although some of these operations necessitate extra attachments. Fig 184 illustrates some of the alternative arrangements possible, and their functions.

Simplicity of design, and the minimum of maintenance requirements, are attractive features of this type of machine, which is

268

184 Swath treatments with a rake-wheel machine. *a* Side raking,
b Gathering, *c* Turning

devoid of gear, chain or belt drives. However, its action is not so
positive as that of some power-driven machines of alternative
design—particularly in heavy crops.

Attachment to the tractor by the three-point linkage is usual,
but some models are marketed with adjustable land wheels to
stabilise the machine in work and for transport. Some front-
mounted types are available, and these are particularly favoured
on farms where peas are grown. Plate 12 shows a rake-wheel
machine swath-turning.

OSCILLATING-RAKE TYPE

The oscillating-rake type of machine (Fig 185) has been extensively
used for many years, although it is now meeting strong competition
from the earlier-mentioned machines. The many available makes
of this machine indicate its efficiency of operation, although it
has many moving parts and is rather cumbersome.

A series of tined rake bars are supported between two *fliers*, i.e.

269

two flywheel-like rotors, one of which is driven by the land wheels. It will be seen that the fliers are arranged obliquely, so that the rakes take on a side-sweeping action as the assembly rotates.

The fliers can be rotated in either direction, so that the machine

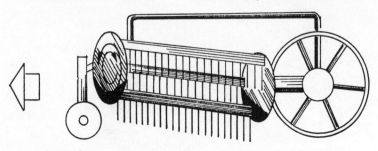

185 Oscillating-rake type machine

can be used for swath-turning with either a forward-throwing or back-kicking action according to the severity of treatment required, for tedding and for side-raking. For turning and tedding, quickly detachable centre sections must be removed from each rake bar, which is wide enough to work two 5 ft swaths.

SPINNING-HEAD TYPE

This is a machine with well-proven characteristics that has lost ground to recent innovations. Its general layout, as seen in Fig 186,

GEAR BOX

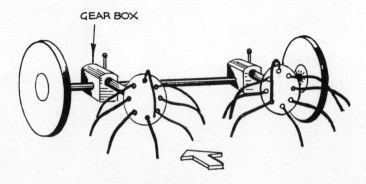

186 Spinning-head type machine

13. Baler

14. Combine harvester

15. Potato digger

16. Sugar-beet harvester with top-saving attachment

is fundamentally simple. The mechanism of the two spinning heads is so designed that the fork-like tines keep pointing downwards while they rotate. Direction of rotation can be changed on both heads, and screens or deflector boards fitted, so that swaths may be turned or collected, usually two swaths combined in one windrow.

The drive for the mechanism is generally provided by the land wheels—even on mounted-type machines. The spinning heads can be raised, lowered and adjusted laterally to match the spacing and width of the swaths.

RAKE-CHAIN TYPE

These machines employ a p.t.o. driven, chain-type cross-conveyor as shown in Fig 187. All the standard haymaking operations can

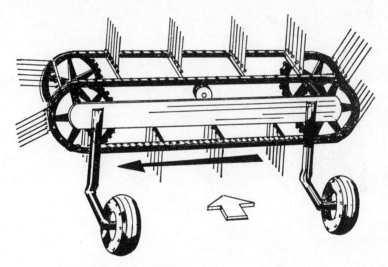

187 Rake-chain type machine

be performed without altering the machine's right-angled disposition in relation to the direction of travel. Although it is designed for attachment to the three-point linkage, adjustable

K

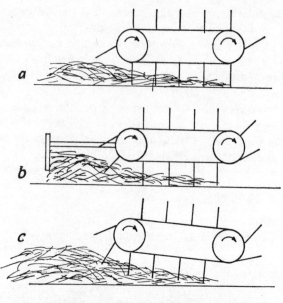

a

b

c

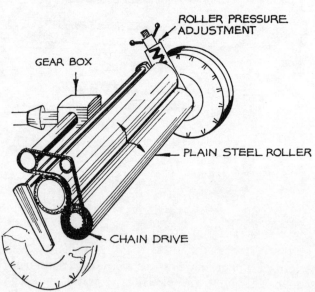

ROLLER PRESSURE ADJUSTMENT

GEAR BOX

PLAIN STEEL ROLLER

CHAIN DRIVE

land wheels are provided. These wheels gauge the general height of the machine over land undulations, and provide the main settings for the machine's various operations. Reference to Fig 188 will show that the various actions are achieved simply by tilting the machine or altering the chain speed. The positions for tedding and spreading are identical. For spreading, however, increased chain speed relative to the ground speed causes the swath to be thrown further, and so the ratio of ground speed to p.t.o. speed is an important factor in the machine's performance. For turning and side-raking, the rake chain is level, but for side-raking, a screen board checks the material and forms a high and narrow windrow. In any operation, the machine is not unduly harsh in its movement of the crop, and high working speeds are possible.

The absence of rotary parts handling the crop eliminates any tendency of the hay to wrap or rope. As with certain other types of hay machine, this one can be drawn sideways to reduce width for transport.

CONDITIONERS (CRIMPERS AND CRUSHERS)

Reference has been made to the reason for crimping and crushing crops intended for hay. Fig 189 illustrates a crusher—the distinction between this and a crimper being plain rollers instead of fluted. Various roller patterns are used by different manufacturers, some being fabricated from strips of steel, while others possess a grooved rubber surface. Fig 190 shows the action of a crusher and a crimper respectively. Fig 191 shows a combination of fluted steel and plain rubber rollers. The advantage of rubber over steel rollers can be seen in Fig 192 which shows the ability of the rollers to accommodate lumps of material without being forced apart. One or both of the rollers may be driven by the tractor p.t.o., and a slip clutch is incorporated in the drive. Some models also

---

188 Rake-chain swath treatments: *a* Turning, *b* Side-raking (note screen), *c* Tedding and with increased chain-speed spreading.
189 Hay conditioner with plain surfaced rollers for crushing.

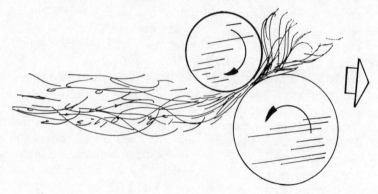

a Crusher

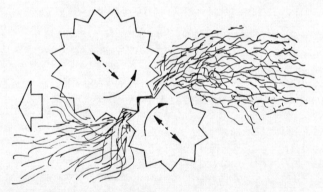

b Crimper

190 Hay conditioning processes: *a* The smooth rollers split the stems of the crop lengthwise. *b* The fluted rollers crimp and bruise the stems at close intervals

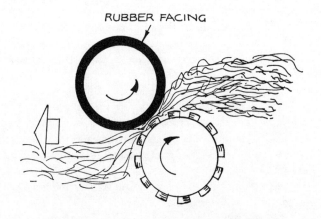

RUBBER FACING

191 Combination of rubber and fluted steel rollers

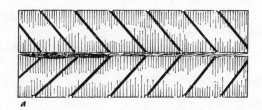

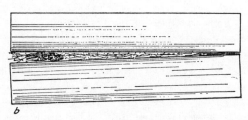

192 Rubber-faced rollers (*a*) accommodate wad of hay; solid rollers (*b*) are forced apart reducing effect on rest of hay

have an over-run clutch. When both rollers are driven, there is less chance of slip or friction on the crop passing through.

As mentioned previously, crimping should be performed immediately behind the mower, that is while the stems are still firm. If the material is allowed to wilt before it is crimped, the operation is less effective and the crop tends to wrap round the rollers. Good results have been achieved by combining the mowing and crimping operation on one tractor using a side-attached mower. The problem of cornering, in this case, may be overcome by first clearing a headland with a forage harvester, and then working in lands.

Some manufacturers are developing self-propelled combined mower/conditioner machines.

Ground speed must depend on the size of the swath, and as the rollers are power-driven, careful tractor gear selection is necessary to ensure that the peripheral speed of the rollers exceeds the ground speed of the outfit so that the crop passes through the rollers in a thin train. Roller pressure on the material can be regulated by varying the pressure spring tension, and this must be adjusted according to the condition and bulk of the swath, e.g. succulent, leguminous crops generally require less pressure than a more fibrous swath of meadow grass.

When the swath size permits a high working speed, the crimper has a tedding effect on the discharged material. Where side deflectors are fitted to the machine, this effect can be exploited to produce a high, well-aerated windrow. Most crimpers are designed for single-swath operation, and the rollers should be set as high as clean pick-up of the swath will allow. On temporary leys, this is particularly important because of the greater risk of stones being taken into the machine. If this happens, the roller drive is usually disengaged by the slip clutch or protective shear-pin.

The crimper is an asset when used in conjunction with barn hay drying, and a further application is in silage making where the crop has been cut with a mower. Crimping of the green material accelerates reduction of excess moisture, so providing a good quality material for ensiling. This reduces both nutrient losses and effluent from the clamp or silo.

276

Field procedure

The operation of swath-treatment machinery has been fairly well covered, but it is worth recapping points of special importance. These are:

1. Avoid excessively harsh treatment of the crop—especially in the later stages of drying.
2. Where possible, avoid wheel marks on the crop.
3. Regulate speed and setting to avoid roping of the crop.
4. Aim at even and consistent windrows.
5. Where possible, follow the direction of the mower.
6. Keep shafts clear of entwined hay.

Maintenance

The lubrication requirements of the various types of machine in current use vary with the type of bearings fitted. Some have totally enclosed ball or roller bearings, some plain bronze bushes and others nylon bushes. Also, gear and chain drives may be enclosed in an oil-bath or exposed and designed to run with the minimum of lubrication. The lubrication of any machine should be carried out as specified by the maker.

Tines, where they are fitted, should be frequently checked for tightness. A spring-steel tine that comes adrift will affect the efficiency of the machine, and can cause serious damage if picked up by a baler.

Belt drives, slip clutches and tyres (where applicable) are other items needing periodic attention. The short time it takes to coat the tines, stripper bars and canopies of these machines with rust preventive after use is always well repaid when the machine is next required, by eliminating the tendency of the crop to hang on these surfaces.

REPLACEMENT PARTS

Spares held on the farm for this range of equipment are usually

277

confined to tines and their retaining bolts, and rubber belts where appropriate.

Safety precautions

The hazards in the use of haymaking machinery are similar to those outlined for cultivating machinery and reference should be made to the safety precautions at the end of Chapter 2.

9 BALERS

The baler is often considered a relatively recent innovation, but the idea of compressing bulky farm materials into bundles or bales is not new. However, the study of early presses and balers is now of little more than historical interest, and this chapter concentrates on the mechanical principles and operating techniques of automatic twine-tying pick-up balers.

A brief reference may first be made to the stationary baler, which gained considerable popularity in Britain during and immediately following World War II, and may be credited with the introduction to many farmers of the benefits of baling hay and straw.

The stationary baler was heavy and cumbersome. Generally, two men were required for the tying operation which was normally performed with wire, although machines could be adapted to take a semi-automatic twine-tying device which required only one operator. Most stationary balers were designed primarily for baling straw behind a threshing machine, and so the feeding trough was high off the ground. When they were used for hay baling in the field therefore, much effort was required to feed the machine. Another drawback of the stationary baler was the absence of a shearing knife on the plunger, which resulted in the bales being formed as an almost unbroken concertina of hay, making it difficult to shake out and ration to stock without waste.

The advantages of the pick-up baler, with its ability to collect the crop from the swath and to tie the bales automatically, were obvious on its first appearance. In a very short time, many farmers who were not at first enthusiastic succumbed to this revolutionary method of hay collection. Packed in bales, the hay could be conveniently brought to the farmstead, so eliminating the laborious business of cutting the hay from stacks and hauling from remote fields. It also simplified the problem of straw collection behind the combine harvester. Plate 13 shows a baler at work in combine straw.

Applications of the pick-up baler

The main applications of the pick-up baler are the baling of hay from the swath and the baling of straw discharged from the combine harvester. Some farmers experimented with the baling of green crops for silage but this never became very popular. Occasionally pick-up balers are used for the stationary baling of hay and straw.

These applications are aimed at simplifying the handling, transport, storage and ultimate use of the material. A further important advantage of baling hay, is the reduction of weather damage risk, as baling can commence before the hay would be dry enough for stacking in the traditional way.

The moisture content (M.C.) of hay to be stacked must be reduced from its 80% or so to about 20%, whereas the acceptable M.C. for baling may be as high as 35%. The optimum M.C. will depend upon the nature of the herbage and the type of baler to be used. The difference between these acceptable limits of moisture content could represent an extra day's drying in the windrow, and the extra chance of damage by rain. For barn or tunnel drying, hay may be baled at a M.C. of up to 60%.

The baler is sometimes blamed for deterioration of the crop by the stripping off or shattering of the protein-rich leaf by its feeding, packing and compressing processes. Whilst there are some grounds for this, it is true to say that more hay deterioration is caused by untimely and inappropriate use of swath-treatment machinery than by the baler. Competent use of the baler will minimise hay damage, and the principles of good operating technique are dealt with later.

Types of baler

Like most other farm machines, balers may be classified in various ways, such as by bale density range (medium density or low density); crop intake to bale chamber (side-feed, overhead-feed, or direct [front]-feed); system of bale formation (plunger ram, press or forming belts); source of power (tractor p.t.o. or auxiliary engine).

Self-propelled balers have been produced but these met with very little enthusiasm.

It may only confuse the uninitiated to define all the above characteristics in an abstract manner at this stage. The various terms will therefore be explained whilst the construction of balers is considered under the broad headings of *medium-density* and *low-density* types. A type of baler capable of wide bale density variation, and which produces unique cylindrical bales, will also be outlined. Some automatic wire-tying pick-up balers and most of the now obsolescent stationary balers, may be classified as *high-density* machines, but the small number of these in current use hardly justifies detailed description.

Construction and working principles of medium-density plunger-type balers

Medium-density plunger-type balers produce bales about 14 in by 18 in in cross-section, of variable length and with a density range of from 5 lb to 15 lb per cubic foot depending on the nature and M.C. of the crop. One distinguishing feature between various makes and models of this type of machine is the direction of the crop intake to the bale chamber, i.e. side-feed or overhead-feed. The majority of balers in current production are of the side-feed type.

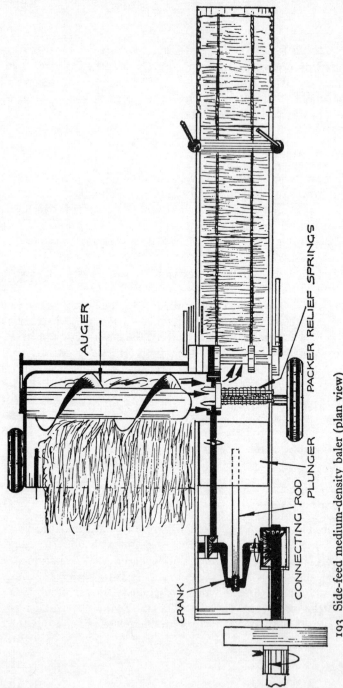

AUGER

CRANK

CONNECTING ROD

PLUNGER

PACKER RELIEF SPRINGS

193 Side-feed medium-density baler (plan view)

SIDE-FEED

Figs 193 to 195 illustrate a typical side-feed baler at work and identify its various components, the functions of which are as follows:

Pick-up mechanism

The pick-up usually consists of a series of revolving tines which retract under the control of cams. The tines rotate in the opposite direction to the land wheels, and so lift the hay from the swath or windrow on to the baler intake table or trough, and into contact with an auger or tined cross-feed conveyor which carries it towards an aperture at the side of the bale chamber (Fig 194). The height of the pick-up above the ground is adjustable. A spring-loaded sheet-metal or tined crop-guard above the pick-up ensures a positive purchase on the crop. Its presence is particularly important in windy conditions, and on some machines its pressure on the crop is adjustable.

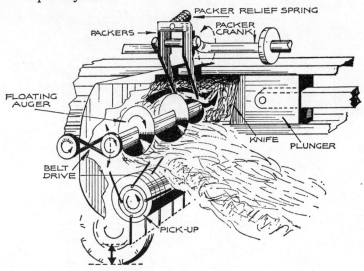

194 Crop intake on side-feed baler

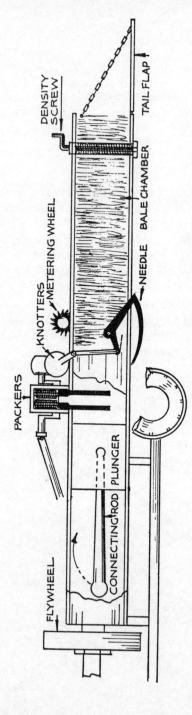

195 Side-feed medium-density baler (side elevation)

Cross-conveyor

The cross-conveyor may take the form of an *auger* or rake, which revolves or oscillates at a speed compatible with that of the pick-up. The auger is belt or chain driven, and permitted to float on the incoming crop, that is it can, within limits, rise and fall according to the bulk of the windrow being handled.

Packer mechanism

A packer mechanism is fitted, where an auger-type cross-conveyor is used, to convey the crop from the end of the auger into the bale chamber through its side aperture—hence the term *side feed* for this type of machine, Fig 194. The packer tines oscillate in synchronisation with the action of the plunger, as they have to sweep across the path of the latter. Certain adjustments are possible on the packers to suit varying crop conditions, and a relief spring or shear-pin is also provided to protect the mechanism from overload damage.

Plunger

The plunger is driven by a connecting rod actuated by a long throw crank-shaft or crank-wheel, Fig 193. The crank may be chain or gear driven from a power-drive shaft or belt driven from an auxiliary engine. The provision of an over-run clutch in the main drive of p.t.o.-driven machines simplifies tractor gear-changing. The direction of rotation of the crank is such that the connecting rod exerts slight downward pressure on the plunger when the latter is on its compressing stroke. A shearing knife fitted to the side of the plunger works in conjunction with a stationary shear plate, on the aperture side of the chamber, to shear off each wad of material fed in by the packer, Fig 196. For efficient shearing action, the clearance between the plunger knife and the shear plate must not exceed the amount specified. Adjustment for this is provided on the rollers, blocks or guide rails on

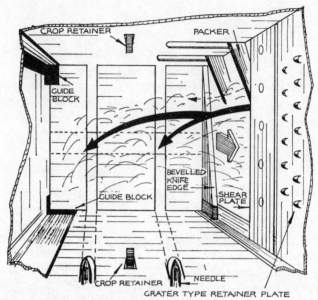

196 View of plunger face, plunger knife and crop retainers inside bale chamber

which the plunger runs. The plunger has two vertical slots through which the needles pass on each tying cycle.

Bale chamber

The bale chamber is the basis of the machine, and determines the width and depth of the bales. Compaction of the hay or straw is achieved by restricting the outlet of the material at the discharge end of the chamber by means of adjustable pressure pads or by squeezing together the side panels, or upper and lower panels of the chamber. The adjustment is usually effected by tension screws, although hydraulic control has been introduced, and is claimed to give more consistent control. A mechanical compensating device is incorporated on some machines to keep the bale density uniform.

One manufacturer provides a device in the bale chamber which gouges grooves in the bale faces to accommodate the bands of twine. This is particularly valuable when baling at high moisture content for barn drying, as subsequent drying causes shrinkage of the hay and a consequent slackening of the twine bands.

The basic depth and width of the bale chamber are not adjustable but, as mentioned previously, these dimensions vary slightly between one machine and another. Most machines produce bales with an end section about 14 × 18 in. The length of the bale is regulable either in *steps*, or by an infinitely variable adjustment.

One make of baler has its chamber arranged cross-wise (Fig 197). This simplifies the intake of the crop and minimises its

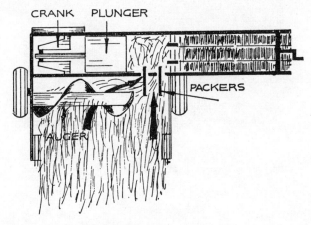

197 Baler with lateral chamber

mechanical handling. Other advantages of this arrangement are the elimination of the forward and backward surging experienced with balers having an in-line plunger action, and good visibility of the bales on discharge. A section of the bale chamber can be folded to reduce transport width.

Crop retainers

The springy nature of the compressed crop necessitates devices to hold the compacted material in position during the return stroke of the plunger—the retainers. These are spring-loaded pivoted claws fitted at the top and bottom of the bale chamber (Fig 198). When the plunger reaches the end of its compression stroke, the newly delivered wad of material is held in place by these retainers.

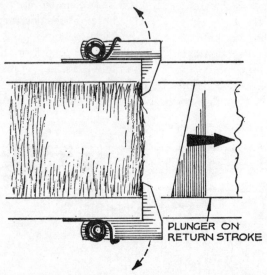

PLUNGER ON RETURN STROKE

198 Spring-loaded crop retainers—holding compressed crop while plunger returns

Sometimes 'cheese-grater'-type side panels are also fitted on each side of the chamber to help hold the crop (Fig 196). The efficient functioning of the retainers is most important to prevent the crop springing back and obstructing the path of the needles.

Tying mechanism

Those who are familiar with the long-established binder knotter, will not find it difficult to follow the principles of the baler knotter, but a simple description of the tying mechanism should be of interest to the less experienced operator.

First, it is necessary to know the names of the principal components and their respective locations. These are illustrated in Figs 195 and 199.

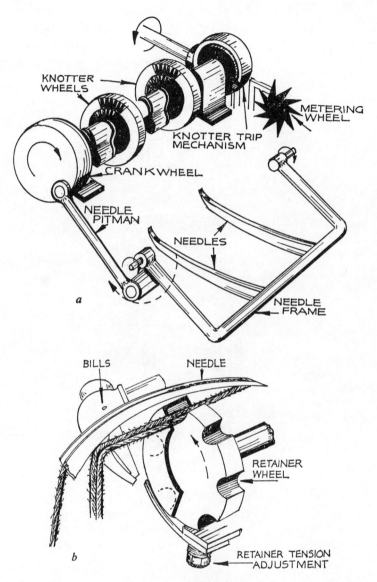

199 *a* Knotter-wheel and needle drive system. *b* Retainer gripping one end of twine; other end is placed in groove and about to be gripped by rotating retainer wheel

Fig 200 shows the path of the twine from the ball to the knotter, when the machine is threaded and prepared for a normal working cycle. Fig 200a shows the twine held at the knotter by a *twine*

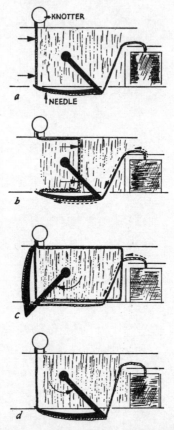

200 Path of twine from ball to knotter: *a* Needle at rest position, *b* Bale moving twine along chamber, drawing twine from ball, *c* Needle placing twine on knotter bills, *d* needle returned

retainer. Fig 200b shows the twine in the bale chamber moving rightwards as the new bale is being formed. When the bale has reached its predetermined length, the needle swings about its pivot, upwards through the chamber, until it reaches the knotter at the top. At this point the needles lay the twine on the *knotter bills* and into a recess in the *retainer* (Fig 199b), which is precisely positioned to receive it, and rotates to grip it firmly. In effect the needle has now threaded two lengths of twine through the bale chamber to the knotter. Note—there is now a length of twine on the inside and outside of the needle quadrant —one to complete the band around the formed bale, and the other to commence the formation of a new band for the bale which is to follow. The knotter bills (which consist of a pair of jaws at the end of a spindle) now rotate and form a double overhand knot to secure the band ·on the completed bale (Fig 201). When the knot has been formed on the bills, the completed band is

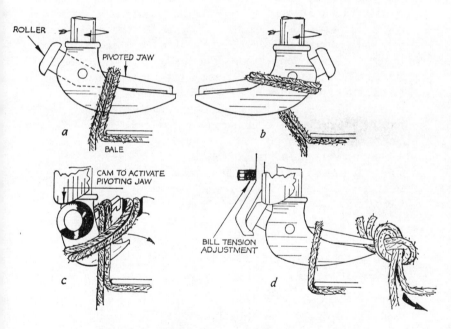

201 Action of knotter bills: *a* Twine on bills at start of turn, *b* Twine wrapped around bills, *c* Pivoted jaw opens to clasp twine, *d* Bills close and grip twine. Ends severed from retainer by knife *not shown*. Outgoing bale pulls knot tight and off the bills

severed from the retainer by a knife situated between the retainer and the knotter bills. As the completed bale passes on along the bale chamber, the loop of the knot is drawn over the ends of the twine which are held in the spring-loaded jaws, until the knot has been pulled tight.

From this description, it will be clear that accurate timing of all the components is vital. In order to synchronise the timing of the needle with that of the retainer and knotter bills, all three components are driven by a common shaft—the needle being actuated by a pitman connected to the shaft, Fig 199a. However, the timing may be disturbed during the life of a machine, and adjustment is provided to correct discrepancies which may arise due to distortion of the needles or to wear on other parts.

The drive to the knotter shaft and needles is brought into operation by a trip mechanism which is released by a metering wheel (Fig 195). The wheel is mounted on the bale chamber so that its star-like teeth project into the chamber to engage on the material moving along inside. The bale-length adjustment is made on this metering unit. Fig 202 shows a typical arrangement for the intermittent engagement of the drive to the tying mechanism and indicates the bale-length adjustment provision. A bale recording meter is usually inter-connected with the tying mechanism to register the number of bales produced.

On most knotters, adjustment of the spring tension on both the knotter bills and the twine retainer is possible, but these should be reset only by someone who fully understands the workings of the mechanism—see also under *Maintenance*.

The gauge of twine most commonly used in this type of machine is that which gives a run of 225 ft/lb.

Mechanical protective devices

Many of the baler's principal components are precisely timed to work in conjunction one with another. Because of this, unless some special provision is made, a relatively small obstruction or mechanical defect could cause a chain of events resulting in serious damage to many parts of the baler. Such provision is made in the form of shear-pins and slip clutches on many of the machine's

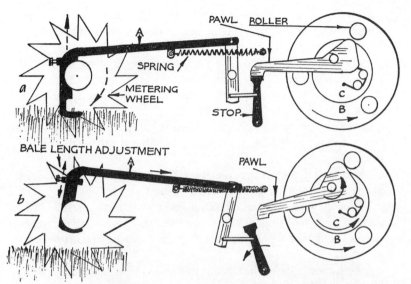

202 Knotter trip mechanism: *a* Pawl resting on stop. Flange B is
constantly rotating, hub C is static. Metering wheel is rotating and
lifting link A. *b* Link A, at top of travel, springs rightwards (by
virtue of recessed tail-piece), withdrawing pawl stop. Heel of pawl
now driven by roller on flange B, causing hub C to rotate and
actuate needles and knotters. Bale length is set by the vertical
position of the collar on link A

components. Examples of these follow together with their possible
locations:

1. A friction-type main drive clutch is usually incorporated
in p.t.o.-driven machines. This cushions any sudden torque
loading on either the baler mechanism or the tractor p.t.o.
system.

2. A friction or serrated-plate slip clutch is usually incorporated
in the pick-up drive.

3. On a few machines, a slip clutch or shear-pin is incorporated
in the auger drive.

Note: The effectiveness of slip clutches relies on their condition
and adjustment. They should be set to transmit normal operational
torque but to slip at any provocation beyond this. This is parti-

cularly important when a machine starts work after winter storage. The safety factor in the use of shear bolts and pins relies on their shear strength being of a known value, so they must always be replaced by bolts or pins of the specified type. Some of these should always be kept on hand when working in the field.

4. Shear bolt(s) are fitted between the flywheel and the main drive or crank-shaft.

5. Shear bolts are sometimes situated in the packer drive system, together with a relief spring to absorb undue strain on the packer tines.

6. A shear bolt and/or break-away device is generally incorporated in the needle-actuating mechanism.

7. A safety stop obstructs the plunger if the needles jam in their path. The severe impact of this sudden stopping of the plunger shears the flywheel bolt(s) referred to in 4 but no serious harm is suffered.

The importance of the needle-protection devices referred to in 6 and 7 justifies more detail.

If for any reason the needles are obstructed in their cycle of operation, the drive is disengaged by means of shear-pins or break-away devices, to protect both the needles themselves and their actuating mechanism from damage. This, however, may leave the needles stationary within the bale chamber, and in the path of the plunger, so the plunger stop is necessary to protect them from serious damage. The shear bolt between the flywheel and crank-shaft is severed by the sudden stopping of the plunger, but this is a small price to pay for the protection of the needles. Before a new shear bolt is fitted and the needle drive reconnected, the cause of the obstruction must be ascertained. This may necessitate clearing the bale chamber completely, and turning the machine over by hand to check the sequence of operations.

8. Shear pin(s) are commonly fitted in the knotter-drive mechanism.

9. An emergency twine cutter is fitted on some balers between the twine containers and the needles. In the event of excessive tension on the twine resulting from tangling at the ball, the twine is severed to protect the needles from undue strain (Fig 203).

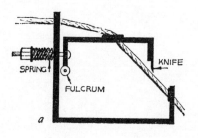

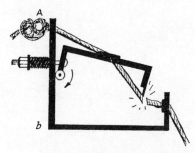

203 Twine cutter: *a* Normal tension: knife clear of twine, *b*
Excessive tension caused by tangle at guide A; pulls knife down
to cut twine and protect needles from strain

OVERHEAD FEED

Although now somewhat obsolescent, the overhead-feed baler
justifies mention since a few models are still in production, and
many are still in use.

On the overhead machine, the pick-up is of a similar design to
that of the side-feed model, but when the crop has been collected
it is immediately raised above the level of the bale chamber and
conveyed crosswise to the open top of the chamber. On each
return stroke of the plunger, a vertically acting packer pushes the
crop downwards into the chamber, hence the term *overhead feed*
(Fig 204).

As the crop is fed in at the top of the chamber, the knife has to
be situated on the upper face of the plunger instead of on the side.

295

As the plunger has its knife fitted to its upper face, the needle clearance slots are arranged horizontally through the plunger. This means that the needles must be positioned on one side of the bale chamber to swing horizontally through the chamber to the knotters on the other side.

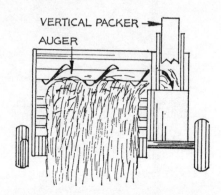

204 Overhead-feed baler (front elevation)

This type of machine produces bales of similar dimensions to the side-feed type, and the same gauge of twine, that is, 225 ft/lb, is used. Very high baling rates are possible with some overhead-feed machines, but the additional elevator, and more elaborate packing mechanism necessitated by the overhead feed, means more weight and extra maintenance. Also, the elevating action subjects the crop to undesirable additional mechanical handling. For these reasons it has been superseded by side-feed types.

Attachment of baler to tractor and preparation for work

Balers are generally fitted with a swinging, or otherwise adjustable drawbar to reduce the overall width of the outfit for transport. On p.t.o.-driven machines, the maker's specifications for the drawbar position and power-drive shaft dimensions should be complied with, to ensure the necessary flexibility—vertically, laterally, and longitudinally. It is most important that guards are fixed in place immediately preliminary servicing has been carried out. Where an auxiliary engine is fitted, its lubricating-oil level should be checked, its air cleaner serviced and its fuel tank filled. It is also advisable to run the engine light for a short time and to check on oil pressure where this is possible. When the drive is eventually engaged, either by belt tension or clutch, this should be done gradually.

Assuming the machine to be in good mechanical condition, it must now be lubricated systematically according to recommendations, the tyre pressures must be checked (the two tyres of the baler are usually of different sizes and require different pressures). The tying mechanism must be threaded with the correct grade of twine. The twine containers usually hold at least four balls, although of course, the twine is only drawn from two of them at a time. The twine must always be drawn from the centre of the ball; thus in order to ensure continuity of the supply, the inner end of each spare ball must be joined to the outer end of the preceding ball (Fig 205). In joining the balls the end should be tied together with the *square* knot shown and any surplus ends cut off. To avoid tangles in the twine, a guide is usually positioned immediately above the centre of each ball and this should not be overlooked when threading. From this central guide, the twine passes through a tensioner which should be set to impose a

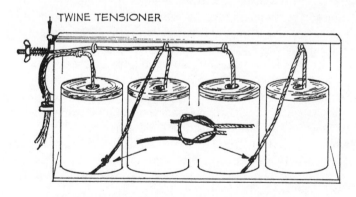

205 Twine box with four balls connected in two pairs. Note square knot

resistance to the twine withdrawal of about 5–7 lb. From the tensioner, the twine passes on to the needles via a number of guides. The needles may be of either channel or tubular construction, and the twine passes either round or through these to the eyes. For initial threading, the ends of the twine are drawn through the eyes of the needles and secured to a convenient rearward point on the baler frame. It is then only necessary to trip the tying mechanism with the baler running for the needles to complete the threading operation automatically.

Before work begins, the bale recorder should be set to zero, and the bale chamber tail flap should be lowered to its horizontal position.

OPERATIONAL ADJUSTMENTS

By far the greater number of balers are of the medium density plunger type—all of which have comparable adjustments—and so description of operational adjustments is confined to such machines.

Pick-up

The pick-up is usually set by a manual lever, although some machines have remote hydraulic control. The tines should be set as high as possible, compatible with clean picking-up of the swath. This avoids undue strain on the tines, and the risk of collecting stones. Some pick-up assemblies are counter-balanced to give a better floating action, and to relieve the impact inevitably imposed on the tines when working on very uneven land. On some balers the pick-up is positively driven from the machine's power source at all times, while on others an optional alternative drive can be obtained from a land wheel of the machine, so keeping the pick-up speed in a constant ratio with the ground speed. This land-wheel drive is generally kinder to the crop, but less efficient when tackling heavy or badly weathered windrows.

The crop guard lying on top of the pick-up should be set according to the volume of the crop.

Auger

Not all machines employ an auger-type cross-conveyor, but where it is used, a variable tension balance spring is usually fitted to assist its floating action on the flow of material entering the machine. The minimum auger weight for satisfactory feeding should always be aimed at—especially with *brittle* hay.

Packer mechanism

The packers are responsible for gathering the crop into the bale chamber. Whether the machine is of the side-feed or overhead-feed type, adjustment is always provided so that the penetration of the packer tines or head into the bale chamber can be regulated. In average working conditions, the degree of packer penetration is not particularly critical, but when baling crops which are either soft and limp or light and *springy*, it can seriously affect the evenness of packing and produce badly shaped bales. Fig 206 illustrates

206 Curved bale resulting from uneven packing

a curved bale which results when more material is packed into one side than into the other. The actual adjustment may involve either altering the length of the packer's effective stroke or adjustment of the packer relief spring tension (Fig 193). The appropriate instruction book should be consulted for details of these settings.

Plunger

The only adjustments provided on the plunger are of a main-

299

tenance rather than an operational character, and these will be dealt with in the appropriate section, but attention must be given to its speed of operation. This is often about 60 to 70 strokes per minute, and the precise speed specified by the makers must be maintained. Even a few strokes per minute below the recommended speed can seriously affect output. Plunger speed on engine-driven balers is of course controlled by the speed of the auxiliary engine and on p.t.o.-driven models by the speed of the tractor engine. In either case, it relies on an efficient engine governor.

Bale density

On some machines, bale density is regulated by reducing the dimensions of the bale chamber at its discharge end by means of spring-loaded hand screws. On others, the method is refined by the use of adjustable pressure pads. These may be set by hand screws or by a hydraulic ram. The latter, not being influenced by the varying tensions of springs, gives greater consistency. This adjustment is influenced by the type and condition of the material being baled. Needless to say, the resistance to the passage of the material through the chamber is simply frictional, so that crops with a high moisture content tend to bind in the chamber, and the tension screws need to be slackened to avoid excessive compaction of the material and overloading of the baler mechanism. At the other end of the scale, dry slippery material requires increased resistance in order to produce a firm bale, and it is sometimes necessary to fit wedge-like cleats inside the bale chamber to produce bales of the required density (Fig 207).

In any baling operation, frequent checks must be made on the weight and tightness of the bales being produced; indeed, with hay baling this is vital, as the amount of moisture on the surface of hay can vary within an hour or so to such a degree as to completely upset the density setting. Excessive compaction of hay can have a number of detrimental effects—loss of leaf due to the extra impact of the plunger and the higher number of strokes per bale, and deterioration due to moulding, especially where the moisture content is above the optimum level. Other undesirable

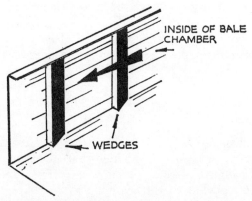

207 Bale chamber with wedges to increase density

features resulting from this maladjustment are excessively heavy bales (which may burst), and faulty tying due to the twine being pinched between two consecutive bales, and, as a result, being pulled out of the retainer (Fig 208). Recently introduced automatic compensating devices make slight modifications to density adjustment necessary. It should perhaps be mentioned here that ground speed of the baler can influence the density of the bales produced, but more will be said about this under *Field procedure*.

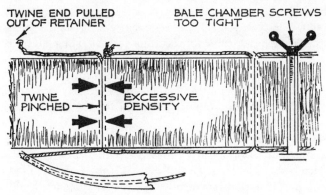

208 Excessive tension of chamber screws prevents twine from drawing between bales

Optimum bale density depends upon the moisture content of the crop. A density of approximately 10 lb per cubic foot with hay of 25% moisture content gives a firm bale without risk of moulding. For high moisture contents, the density should be reduced.

Bale length

On most balers, this is infinitely adjustable. Fig 202 shows the principles of the mechanism. Although 36 in is a common length, the range of adjustment may permit lengths from 12 to 50 in. When hay is baled at the higher limits of moisture content, a shorter bale length reduces the twine slackness caused by after-shrinkage. This can be considerable, especially when the hay is barn dried.

Working principles of low-density press-type balers

The low-density press-type baler has a wide, short bale chamber, and the crop is comparatively lightly compacted by a swinging plunger which operates with a pendulum action (Fig 209). This action of the plunger is less harsh on the material than is the plunger of the medium-density machine.

Lower density and gentler crop treatment assist the production of better-quality hay. Hay can be baled with a higher moisture content than is practicable for medium-density baling, and there is less risk of crop damage in the form of leaf loss. However, bales produced by this type of machine will absorb more moisture if it should rain between baling and collection, and their looseness and lack of positive shape are less convenient for stacking and subsequent handling.

The crop is picked up and conveyed straight into the front of its in-line bale chamber by a fork or paddle. The bale density is controlled by screws at the end of the chamber, and the tying is performed automatically by two knotters using thinner twine than medium-density machines—usually of the 400 ft/lb gauge.

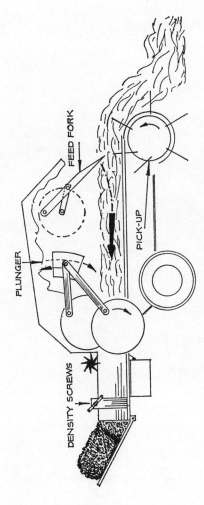

209 Low-density press type baler showing pendulum plunger action and crop flow

FEED FORK

PLUNGER

DENSITY SCREWS

PICK-UP

L

Working principles of roll-type balers

Fig 210 shows a roll-type baler. It became popular some years ago and was so successful that no comparative study of balers would be complete without its inclusion.

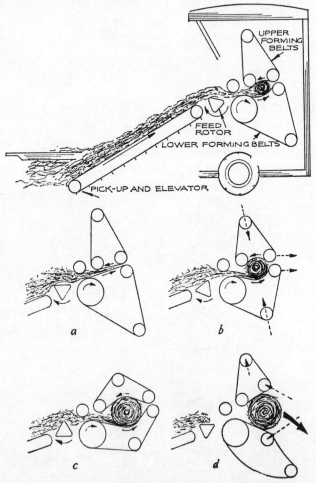

210 Layout of roll type baler: *a–d* Formation and ejection of roll bales

It collects the crop from the swath by means of a chain and slat conveyor which also elevates it to the bale-forming mechanism. This consists of two sets of flat rubber belts each running on three rollers (Fig 210). The train of material is first lightly compressed and then fed into the moving belts in an even flow. The opposing faces of the belts in contact with the material create a rolling action which continues until the cylindrical bale reaches a predetermined size. As its size increases, the outer belt rollers move to provide the necessary increase in effective belt length.

Another unique feature of this machine is the absence of knotters for securing the twine. Instead, the twine is fed into the machine just as the bale is being completed, and an arm guides it across the full width of the belts so that a band of twine is wound spirally around the bale (Fig 211). The number of coils of twine

211 *Top* A well-formed roll type bale. *a* Too narrow a windrow. *b* Inaccurate driving or uneven windrow preparation

wrapped around the bale can be varied to suit the nature of the crop. As the bale is discharged by the parting of the rear belt rollers, the expansion of the hay or straw tightens the twine and holds it in place. The gauge of twine used is usually 500 ft/lb. The

305

diameter and density of the bale can be widely varied and the pressure of rolling can be regulated to give a 'soft-centre' bale to permit baling at relatively high M.C.s. The roll method of bale formation is gentle on the crop, and produces bales that are less absorbent than any other type. A special technique is necessary for successfully stacking the cylindrical bales, and the manufacturers of the machine offer advice on this.

With earlier models, the outfit had to stop as each bale was bound with twine and discharged, but this drawback was later overcome. In any event, the machine's capacity to handle four swaths in one windrow more than compensates for intermittent discharge stops. It is important, however, that the windrows are formed to match the width of the baler's pick-up, and that the outfit is accurately driven or bales shaped like those in Fig 211 may result.

Field procedure

CONDITION OF CROP

For efficient baling, the windrow must be properly prepared. Briefly, this means that it should be of even height and width, of uniform density, and free from the *roping* effects sometimes produced by the incorrect use of swath-treatment machinery. A further factor is the important matter of moisture content, to which reference has already been made.

The number of swaths turned into one windrow depends upon the density of the crop and the intake capacity of the baler.

Normally the baler travels round the field in the same direction as the mower and swath-treatment machines. Sometimes, the corners of the field are windrowed and baled diagonally before the sides are worked—this helps clean pick-up of the crop at the corners. Another method is sometimes used where the crop has been cut and crimped simultaneously by one tractor in lands—in this case the baler also must work in the same sequence. When starting work in a hedged-in field, either in the morning or evening, it may be an advantage to start several windrows in from the field boundary where the crop is generally in a better condition for baling.

DRIVING TECHNIQUE

Before the machine is actually put into work, a check should be made on its running speed. On a plunger-type baler this may be done with the aid of an ordinary watch. The baler should be run, and a count made of the number of plunger compression strokes per minute. If any discrepancy exists between this and the maker's recommendation, the tractor or baler engine speed must be modified. As has been stated, a few strokes per minute below the recommended plunger speed can seriously affect the machine's output. On power-driven machines, correct relationship between p.t.o. speed and ground speed is essential for good work—a multi-speed gear box on the tractor and an over-run clutch on the baler are a great help in matching the outfit to crop conditions. Excessive ground speed can result in insufficient bale density, untidy packing and erratic metering-wheel drive with its consequent ill-effect on bale-length control. In extreme cases, the machine may be completely blocked. Insufficient ground speed, on the other hand, may result in crop damage due to the excessive number of plunger strokes per bale. On most medium-density balers, about 15 strokes of the plunger between bale discharges (36 in length) generally gives satisfactory results, with a little latitude either way according to the moisture content of the crop. At the end of the day, or whenever the machine is shut down, the mechanism should be run long enough to clear all the crop-intake components of hay or straw.

OPERATIONAL CHECKS

The baler's operational adjustments have been discussed, but it it worth summarising the various checks which should be made periodically while work is in progress.

1. Bale density. The importance of this adjustment has been stressed, and an indication of the safe density range for average conditions given. When baling starts, it is well worth the time,

particularly for the less experienced operator, to weigh a bale with a spring balance, calculate its volume and work out the precise density of baling. With experience, rough checks can be made quickly in the field by merely handling the bales and checking the tension on the twine bands.

2. Bale formation. Unequal distribution of the material will result in badly shaped bales. This bad distribution may be due to bad windrow preparation or incorrect packer setting.

3. Slicing of material. One face of the bale—the plunger-knife side—should show evidence of a clean cutting action. Any raggedness here would suggest the need for resharpening the shear blade, or resetting its clearance. A blunt knife can sometimes be detected by the sound and impact of its action. On engine-driven balers, the excessive shock load of a dull knife causes violent oscillations of the engine speed at each stroke of the plunger. Roughness of the cut face of the bale could, however, be caused by an accumulation of crop residue adhering to the inner face of the chamber near the shear plate. Such deformation can also affect the bale length metering-wheel operation. To check and rectify this, the bale chamber must be emptied of material.

4. Tying of bales. Complete mis-tying by one of the knotters is too obvious to go unnoticed for long, but occasionally bales which are apparently tied satisfactorily should be examined to see if the knots are properly formed. Irregularities, such as excessively long ends, large bows or the lack of bows, may be clues to slight faults in a knotter's performance. Because of variations in knotter design, between different makes of machine, such symptoms may imply different faults, and so operators should be guided by the relevant instruction manual. No attempt should be made to explore beyond the operator's province in such matters. This usually includes checks to ensure that the twine is drawing freely without tangles or fraying, that the twine tensioner is correctly set, that the needles and guides are correctly threaded and that the retainer and knotter bills are free from accumulated twine fibre and trash (some balers are fitted with a fan to help keep the knotters free from such obstructions). Reference is made to other possible tying faults under *Maintenance*.

5. Consistency of bale length. One possible cause of irregular bale length has been given in item 3. Other causes may include a metering wheel-spindle seized because of lack of lubrication, a faulty trip mechanism or insecure needle retention at the rest position—the last factor is usually controlled by an adjustable friction-brake device.

Maintenance

DAILY ATTENTION

Daily lubrication of the baler, with the specified types and grades of lubricant, should be systematic. Chain and belt tensions should be checked and adjusted if necessary, and the plunger knife should be examined for sharpness. If sharpening is necessary the knife must be removed from the plunger. Sometimes the knife is reversible to provide two periods of service before sharpening. In any event, it is wise to have a spare knife on hand to alternate with the one in use. A good bench grinder is useful for sharpening plunger knives, but there is the danger of softening the steel by overheating unless great care is taken. It is important also to maintain the correct angle of bevel.

The bale chamber should be checked for crop residues adhering to its inner faces, particularly on the plunger-knife side, as this hinders correct bale formation. The crop retainers which hold the material in its compressed position should also be checked for efficient working.

Usual engine maintenance and tyre-pressure checks should also be carried out.

At the end of each day's work, the bale chamber should be slackened off to allow for the expansion of the bales in the chamber overnight.

PERIODIC ATTENTION

At intervals during the season the clearance between the plunger knife and the shear plate should be checked. This clearance is influenced by the setting of the plunger runners or guides, which are adjustable. All manufacturers specify a clearance which should be maintained—usually it should not exceed $\frac{1}{32}$ in. To make the check, the machine should be turned by hand until the plunger knife is adjacent to the stationary shear plate. The plunger should be pushed away from the shear plate to measure the maximum clearance when any play is taken up. On side-feed balers, the plunger must be pushed horizontally; on overhead-feed balers which have the shearing knife at the top of the plunger, the latter must be forced down to check the maximum clearance. When this check is made, the plunger knife must be parallel to the shear plate. On some machines, adjustment is possible to remedy any discrepancy in this respect by raising or lowering one side of the plunger.

The condition of such items as chains and sprockets should be checked, and the pick-up examined for broken or lost tines. Indeed, the whole machine should be periodically checked for evidence of worn parts, loose bolts, etc. Manufacturers discourage operators from attempting any but the simplest of adjustments to the tying mechanism, and this is in the best interests of baler owners. An outline of the more common faults follows, with some guidance as to what may be causing the trouble.

Common tying faults

1. In Fig 212a illustrates diagrammatically a complete band as it should be formed round the bale. The knot shown has its two ends pulled completely through the loop, although some types of knotter produce a bow-like effect when an extra-long end remains in the loop. In fact, it is always the same basic overhand knot.

2. Fig 212b shows a knotting failure. The so-called *needle end* of the band is frayed and its *retainer end* knotted. This is usually caused by excessive retainer tension which should be set to hold the end firmly and yet yield enough twine to the bills as they turn.

3. In Fig 212c, the so-called *needle end* of the twine is knotted, while the *retainer end* is untied and has a clean cut. This free end has been withdrawn from the retainer for one or more of the following reasons:

   a. Excessive bale density pinching the twine between the bale being formed and the previous one, and resulting in the twine being pulled out of the retainer instead of from the ball.

b. Excessive twine tension caused by wrong setting of the tensioner at the twine box (Fig 205).

c. Insufficient tension on the retainer spring (Fig 199b), resulting in inadequate grip on the held end of twine while the bale is being formed.

d. Accumulation of twine fibre forcing the retainer shoe open.

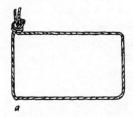

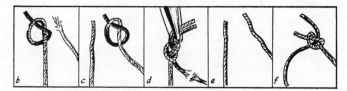

212 Band: *a* Correctly tied, *b–f* Symptoms of irregularities, not necessarily knotter faults

4. Fig 212d shows a correctly formed knot hanging on the knotter bills, and the band itself broken. This may be due to excessive bill tension, rust on the bills, or incorrect position of the stripper arm, where incorporated.

5. Fig 212e illustrates a band which has kinked ends, indicating that a knot had in fact been formed but never completed. This would probably be due to insufficient knotter-bill tension, wear on the bills, or obstruction to the pivoting action of the opening jaw by twine or crop fibre.

6. Fig 212f shows a slip knot. The needle twine is uncut and so remains connected with the twine source. This could be due to

inadequate needle penetration, lateness in the needle timing, or advanced retainer timing.

Most of the above faults can occur on all makes of knotter; other faults are peculiar to particular machines. Worn, damaged or mistimed components could result in twine wrapping round the bills, repeated unthreading of the knotters, or the joining of one bale to another. Investigation into these more subtle problems should be left to the agricultural mechanic. To *discourage* operators from interfering with the tying mechanism, some of the complex mechanical factors involved are listed:

1. Timing of needles in relation to retainer and bills
2. Timing of needles in relation to plunger movement
3. Lateral alignment of needles on side-feed machines, and vertical alignment on overhead-feed machines
4. Position of twine fingers (where fitted)
5. Position of stripper arm in relation to bills
6. Position and condition of twine knife
7. Clearance between the knotter bills when open.

Most of the above components are likely to become disturbed only as a result of some unusual stress or when maintenance is neglected. On rare occasions, damage may be caused through twine tangling and resisting the needle-drive mechanism. It is to avoid this that the twine-cutter protective device (Fig 203) has been introduced. In any event, balls of twine should always be stored on end, and in a dry place to preserve its condition.

Erratic bale length is not, strictly speaking, a fault in the tying mechanism proper, but it is convenient to refer to it at this stage. It may well be due to bad packing of the bale—generally as the result of incorrect ground speed or badly prepared windrows. Sometimes, however, it may be due to a fault in the metering-wheel unit or its related trip-control mechanism, or to incorrect setting of the *needle-retention brakes*—(the friction pads or strips which hold the needles at their rest position).

END-OF-SEASON ATTENTION

Before stowing the machine away, the bale chamber should be emptied, the machine thoroughly cleaned, and a coating of rust preventive applied to all crop-friction surfaces. Steel plunger runners should also be protected from rust.

The whole machine should be checked for damaged or worn parts, and special attention should be given to the condition of bearings, sprockets and chains.

The machine should be lubricated throughout, and if specified by the makers, any oil-bath gear boxes should be drained and refilled with the correct lubricant. The tyres should be relieved of the baler's weight, and covered for protection against oil, grease and daylight. Storage routine should be carried out on the auxiliary engine where this is applicable. On p.t.o.-driven machines, the condition of the power-drive shaft universals should be checked.

If any spares or replacements are required it is advisable to order them at once—quoting the part numbers.

REPLACEMENT PARTS

During the baling season, it is advisable to have the following spares on hand, in so far as they are applicable; pick-up tines, chain links, auger and main-drive belts, plunger knife and bolts, slip-clutch elements and springs, shear bolts and/or pins, as necessary.

Safety precautions

The baler has many moving parts, some of which operate at high speed and come into operation at unpredictable times. Also, it cannot be quickly stopped once its mechanism is in motion. Because of this, great care is needed when operating or carrying out maintenance work on this machine, and the following advice is tendered to all operators:

1. Always use a draw pin of the correct size, and with a safety retaining device.

2. Ensure all guards are securely fixed before work begins, and immediately after making any adjustments necessitating their removal.

3. Never attempt to make adjustments or clear blockages with the

machine running—this is particularly important with regard to the tying mechanism which operates intermittently, but rapidly once tripped.

4. Take special care when removing, sharpening and resetting the plunger knife. The baler flywheel or crank should be locked in some way when carrying out this work. The same precaution should be taken whenever maintenance routine involves putting the hands or arms inside the crank and connecting-rod housing.

5. Before starting the baler mechanism, always ensure that no one is near its moving parts.

At one time, even the most venturesome of British farmers would scorn the suggestion that combine harvesters could be successfully used in this country. The chief obstacles were supposed to be the climate and the size and pattern of our farms. In spite of this prejudice, however, a few machines were eventually imported from America. Some years passed before any real increase in numbers took place, even though some of the pioneers claimed good results from their use.

The combine's saving in manpower as compared with the then conventional methods of grain harvesting was obvious, but cereal growing was not at that time the predominant enterprise on most farms. There was also the question of straw conservation—straw being valued more highly in this country than in the homeland of the combine.

During and after the Second World War, however, the demand for increased output with fewer men, and the need for an increased acreage of cereal crops, forced the farmer to look for quicker methods of harvesting. So the earlier reluctance to accept the combine as a practical proposition was overcome.

The demand for combines soon exceeded the meagre supply available, even though many of these imported models were not entirely suitable for the heavier crops and less favourable weather conditions experienced in this country. However, the early problems were eventually overcome, as is evident from the many reliable machines in current use.

Some credit for the combine's success, in this country at least, must go to the botanists who have done valuable work in introducing cereal varieties suitable for this method of harvesting. Two important crop characteristics achieved are shorter and stronger straw, reducing both the volume of straw to be handled and the risk of 'lodging', and differing periods of maturity which help in the staggering of harvesting operations. Plate 14 shows a combine harvester at work.

Construction and working principles of combine harvesters

A general description of the layout and structure of the combine harvester is best preceded by an analysis of its various functions. Different makes and models of machine vary considerably in size and external appearance, but they all perform much the same basic processes. These are:

1. Cutting the crop
2. Extracting the grain from the ear
3. Separating the grain from the straw
4. Separating trash from the grain
5. (On some machines) screening the grain to give further cleaning and grading of the finished sample

The various approaches of different manufacturers to the achievement of maximum efficiency in these operations has produced a variety of component designs. Before these can be usefully discussed, it is necessary to study the locations and functions of the main components of a typical machine, as shown in Fig 213.

CUTTER BAR

The construction of the cutter bar is similar to that of a mowing machine. Its series of fingers are spaced at 3 in intervals, although on some combines the fingers are longer, as also are the knife

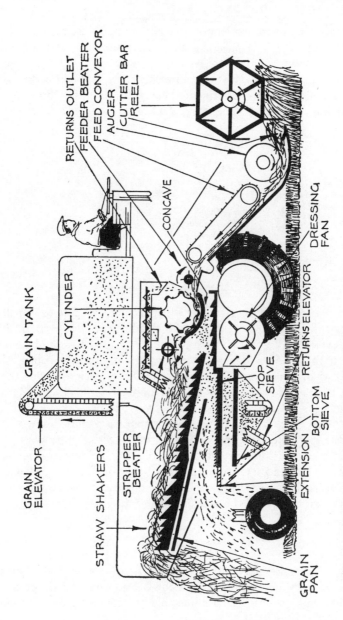

213 Combine harvester layout and crop flow

sections. This increased length of section influences the angle of
the cutting edges, giving a slightly greater shearing, as opposed
to slicing, action. Knives are usually supplied with serrated edge
sections. These cut clean straw more efficiently, and eliminate the
need for sharpening—their action being to perforate the straw
rather than to slice it (Fig 214). Plain sectioned knives, on the

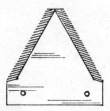

214 Serrated-knife section

other hand, are more effective when tackling a crop in which
under-sown clover has become rather advanced, or green trash
has developed. Some combines have a 3 in knife throw and some
a 6 in throw. However, the significance of this is mechanical
rather than operational. Another knife-throw design moves the
knife a little short of the full 6 in, so eliminating dead movement
of the knife sections at the end of each stroke.

At each side of the cutter bar, triangular panels, known as
dividers, part the crop and determine the effective cutting width
of the machine. Some machines have rotating dividers to simplify
intake in laid and tangled crops. The splay on the dividers may
be expressed as the *gather* because it slightly increases the effective
cutting width beyond the actual width of the cutter bar.

REEL

The reel is positioned above the knife, where it rotates and gathers
the crop into the machine as the cut is made. It is chain or belt
driven, and its position is adjustable to accommodate different
straw lengths and crop conditions.

FEED-TABLE AUGER

The feed-table auger is found only on combines having a cutter-bar width substantially greater than the threshing-cylinder width. The significance of this is discussed later. On self-propelled machines, the auger has left hand and right hand spiral vanes, and these gather the crop from the extremities of the cutter bar, and discharge it on to a centrally disposed feed conveyor by means of retracting fingers. The auger can always be moved vertically, and in some cases forwards and backwards. A chain drive is invariably employed.

FEED CONVEYOR

The feed conveyor carries the crop upwards and backwards to the feeder beater and threshing cylinder. On machines employing an auger, it is usually of the chain and slat type. On machines without an auger, it takes the form of a rubber or canvas conveyor fitted with wood or metal cross slats. In some cases, a trough at the elevator's collection or discharge point acts as a stone trap.

HEADER

The header is the complete assembly which embodies all the above-mentioned components—namely the cutter bar, the reel, the auger (where fitted), and the feed conveyor. The area immediately behind the cutter bar is sometimes called the *feed table*. On most machines, this whole assembly pivots at a point near the threshing cylinder, and can be raised and lowered to regulate the cutting height. Some trailed-type combines employ a fixed header, and the whole machine pivots on its wheel axle as the cutter bar is raised or lowered.

FEEDER BEATER

The feeder beater is positioned between the top of the feed conveyor and the front of the threshing mechanism. It rotates in the same direction as the cylinder, speeding up the flow of the crop and evening out slight irregularities in its volume.

CYLINDER (DRUM) AND CONCAVE

The cylinder and concave are two vital components, comprising the threshing mechanism of the machine, and are sometimes referred to as the 'heart' of the combine. They perform two functions—the extraction of the grain from the ear, and the separation of the grain from the straw.

The cylinder (sometimes called the *drum*) has its periphery made up of a series of beater bars. Various types of beater bar have been used in the past, but the *rasp* type shown in Fig 215 is now in general use.

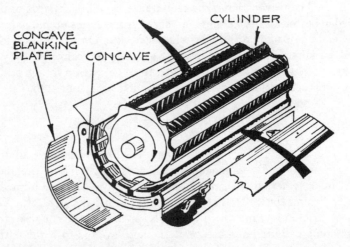

215 Cylinder and concave assembly

The concave, in accordance with its name, is a curved grid also shown in Fig 215. It has to be precisely set relative to the beater bars of the cylinder, which rotates on its central spindle at high speed. This rotation draws the crop in between the beater bars of the cylinder and the cross bars of the concave, and rubs the grain out of the ears in the process. An alternative name for this type of beater bar is the *rub* bar.

Once extracted from the ear, the bulk of the grain passes down through the grid structure of the concave formed by the wire separators, on to the sieves located in the shaker shoe beneath, whilst the straw is directed by a delivery grid on to the *stripper beater*. Needless to say, the cylinder speed and concave/beater-bar clearance are vital factors in the control of threshing intensity. They are discussed in detail under *Operational adjustments*.

STRIPPER BEATER

The stripper beater is similar to the feeder beater described previously. It is behind and slightly above the cylinder, and it controls the transfer of the threshed straw from the concave to the *straw shakers*. Without the beater, the straw would be thrown clear of the front of the shakers and receive inadequate agitation. The beater also helps to prevent straw wrapping around the cylinder—a tendency which can arise in certain crop conditions.

STRAW SHAKER(S)

During the threshing process, some kernels of grain inevitably fail to pass through the concave and are instead discharged along with the straw. To retrieve this grain, the straw is agitated either by a wide, one-piece reciprocating shaker or by a number of narrow shakers working with an oscillating action (Fig 216). The shakers are slatted trays, with steps or projections which lift and jostle the straw. Any loose grain amongst the straw is shaken through to the *grain pan* beneath. From this pan, which also works with a reciprocating motion, this grain joins the main stream of grain which has passed through the concave. The straw shakers are sometimes known as *straw walkers*.

GRAIN PAN

The grain pan is a shallow tray situated beneath the straw shakers to collect the grain which falls through the latter. It is sometimes

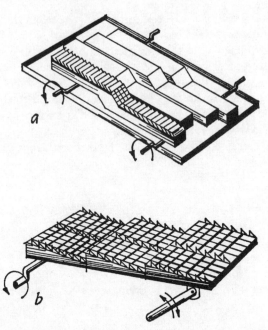

216 Types of straw shaker: *a* Oscillating (box type), *b* Reciprocating (one-piece type)

extended well forward to collect also the main flow of grain from the concave. On some machines, it reciprocates in opposing directions to the *shaker shoe* or to the one-piece straw shaker, to reduce vibration. The pan discharges the grain on to the top sieve.

TOP (CHAFFER) SIEVE

The next stage in the process is the sifting and cleaning of the grain. Although most of the straw will have been diverted, there will still be short lengths of this, together with cavings, dust and possibly other trash, to be segregated from the grain. The differing sizes and weights of these materials enable them to be separated by a sifting and winnowing process.

The top sieve is situated in the shaker shoe, which reciprocates with a short throw. Its upper and lower faces are blown by a current of air from a fan. The grain passes through its openings to the secondary, bottom sieve beneath, while light dust and chaff are blown out at the rear of the machine. Any large, heavy trash present is jogged along the sieve also to be deposited on the ground (Fig 217). The top sieve has several longitudinal ribs on its upper surface to prevent grain sliding down to one side when

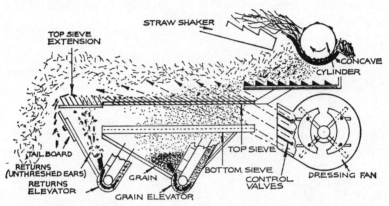

217 Top sieve, bottom sieve, dressing and returns system

the machine is working across a gradient. At the extreme rear of the sieve, an extension riddle or grid is attached to collect any full ears or portions of ears which may have escaped through the concave and missed the threshing process. Further reference is made to this under *Returns system*. Adjustment of the top sieve is considered under *Operational adjustments*.

BOTTOM (CLEANING) SIEVE

The bottom sieve performs the second, and on most machines final, sifting and winnowing of the grain, although some machines do have facilities for further cleaning and grading of the sample. Its action is similar to that of the top sieve, and it is located in the same shaker shoe. This secondary screening, however, requires finer sieve apertures. As with the chaffer sieve, dividing ribs help to keep the grain spread over its surface on hillside work. The

323

bottom sieve may be adjustable or inter-changeable for the harvesting of various-sized cereals.

SHAKER SHOE

The shaker shoe is the framework and housing which embodies the top and bottom sieves. It reciprocates by means of a pitman connected to a crank or eccentric drive.

DRESSING FAN

The dressing fan works in conjunction with the sieves to separate light trash from the grain. The conventional position for the fan is approximately beneath the concave and in front of the shaker shoe. Various adjustments are provided to control the volume and direction of the air stream it delivers.

GRAIN TROUGH AND ELEVATOR

The grain trough is located immediately beneath the bottom sieve to collect the grain. At the bottom of the trough, an auger conveys the grain to one side of the machine and discharges it into the base of an elevator. The elevator—usually of the chain and flight type—delivers the grain to a collecting tank, or to the hopper of a bagging attachment which may be equipped with a final *grading screen*.

RETURNS SYSTEM

The ears of some varieties of wheat and barley have a tendency to break off, and it is possible for these to pass through the concave, and to reach the sieves unthreshed. This tendency can be further aggravated by incorrect concave setting (insufficient clearance at the front), and by the absence of one or more wires which form the grid structure of the concave. Normal sieve settings do not allow such unthreshed heads to pass through to the grain hopper, so without the returns system they would be discharged on to the ground at the rear of the machine. To prevent this, a separate hopper is fitted at the rear of the shaker shoe and beneath the top sieve extension which, as mentioned earlier, has wider apertures than the sieve proper. Any unthreshed ears which are jogged along to the end of the shaker shoe, drop through the extension into the returns trough, from where they are conveyed by auger and elevator to the cylinder for rethreshing.

In addition to retrieving unthreshed ears, the returns system also performs an important function when sieves become blocked —due either to incorrect settings or to damp working conditions. In such circumstances, the overspill of threshed grain from the partially blocked sieves is collected in the system. Needless to say, it is not desirable to subject such grain to a second threshing process, and so some machines have an alternative returns discharge point—usually the forward end of the straw shakers. By this means, the grain is retrieved, but not rethreshed. Clearly, however, unless the cause of the recirculation is corrected, the machine will eventually become choked.

GRAIN DISCHARGE SYSTEM

Certain options in grain discharge are offered with most combine harvesters. One is a simple bagging attachment which comprises a hopper with discharge chutes and a platform for the operator. Where desired, a grading screen can also be provided to segregate trash, broken kernels and immature tailings from the *best* grain to give a good sample (Fig 218). Where bulk handling is required, a large-capacity receiving tank is fitted together with unloading equipment, usually in the form of a spiral elevator (Fig 219). Some machines, however, are now equipped with dual-purpose tanks which may be used for bulk handling *or* bagging-off of the grain.

MECHANICAL PROTECTIVE DEVICES

As with most agricultural machines having components likely to be subjected to excessive strain or impact, the combine harvester has various safety devices incorporated in the drives of its many

325

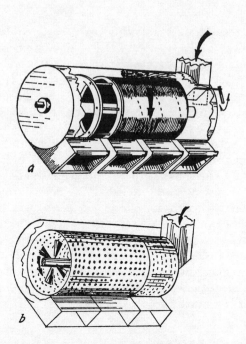

218 Grain-grading screens: *a* Rotating screen with internal spiral vane, *b* Static screen with internal helical rotor

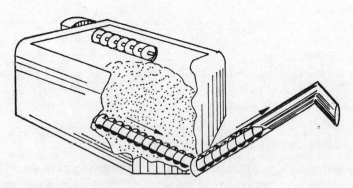

219 Grain tank with spiral discharge elevator

sub-assemblies. Extensive use is made of friction and serrated-plate slip clutches and shear-pins. The following list gives their common locations:

1. Reel drive
2. Knife-drive mechanism
3. Auger drive
4. Feed conveyor
5. Feeder beater and cylinder drive
6. Drive to shaker shoe actuating eccentrics
7. Grain and returns elevators
8. Header-lift mechanism

A further safety device mentioned previously, designed to protect the cylinder and concave from damage, is the *stone-trap* under the feed conveyor.

Types of combine harvester

The broad classifying features to be considered when comparing various types of combines are: the mode of traction, the ratio of cutter-bar width to cylinder width, and the grain discharge system. In making a choice before purchase, there are obviously many more detailed specifications to be studied.

MODE OF TRACTION

The two possible modes of traction are, of course, trailed and self-propelled. Although fewer trailed machines are now marketed, the relative merits of these alternatives are worth considering, for apart from the economic implications, the cylinder width of most trailed machines is approximately equal to the cutter-bar width, whereas that of self-propelled machines is usually considerably narrower. This entails a fundamental difference in crop intake, and the significance of this is discussed later.

The self-propelled machine is more convenient when opening out the field, as it eliminates the back-swath. Also, the weight on its traction wheels helps when working in wet soil conditions and on hilly land, where a tractor-drawn outfit might be troubled with wheel-spin. However, if a large machine is not imperative, the trailed harvester involves less financial outlay and has reduced maintenance costs.

Moreover, recent developments in the form of over-run clutches on the combine, and *live* power take-off and multi-speed gear boxes on tractors, have done much to remove the earlier objections to p.t.o.-driven machines.

CUTTER-BAR WIDTH/CYLINDER-WIDTH RATIO

Fig 220 compares two common arrangements in diagrammatic form: (a) shows an auger-type header which gathers the crop into the centre of the feed table and passes it back into an elevator and cylinder which have a width substantially narrower than that of the cutter bar. Illustration (b) shows the alternative arrangement, used in most trailed machines and in some self-propelled harvesters. In this case the cylinder is approximately the same width as the cutter bar. Combines of the second type are often referred to as *straight-through* machines. Other terms used to distinguish the two layouts illustrated are *narrow-cylinder* type and *wide-cylinder* type, respectively. From the illustrations, it can be appreciated that the cylinder of type (a) has to cope with a much more concentrated flow of material than that of type (b).

Type (a) is generally claimed to damage the grain less by cracking or skinning the kernels during the threshing process, because the bulk of straw passing through provides protection and a slight cushioning effect. The resultant concentration of straw immediately behind the concave, however, is less desirable so far as separation of the threshed grain from the straw is concerned. Oscillating box-type straw shakers are necessary to provide maximum agitation, and to prevent grain being lost with the straw (Fig 220a).

On the wide-cylinder machine type (b), when cutting a normal upstanding crop, the crop is presented ears first to the cylinder in a thin even flow. This makes the grain more vulnerable to beater-bar abrasion, and makes the adjustment of threshing intensity more critical; but, on the credit side, this arrangement improves the separation of grain from straw. A simple one-piece

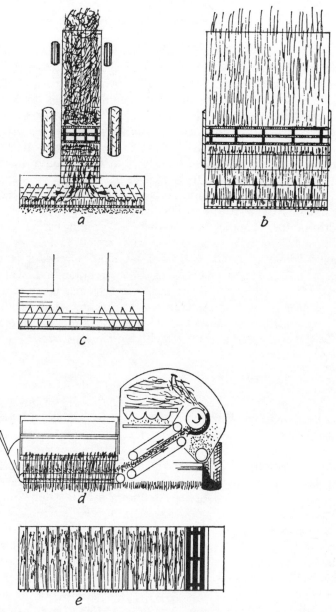

220 Cutter bar/cylinder width ratio and layout: *a* Narrow, *b*
Wide, *c* Intermediate, *d* and *e* Longitudinal cylinder arrangement

reciprocating straw shaker is adequate in wide-cylinder combines (Fig 220b).

Therefore, when operating the narrow-cylinder machine, a close watch should be kept on ground speed and other factors which can cause grain to be lost with the straw. On the wide-cylinder machine, special attention should be given to factors affecting the severity of the threshing process in order to avoid grain damage which may spoil a sample otherwise suitable for seed or malting. Mechanical damage to the grain can impair its germinative energy, and also give rise to deterioration in storage, even when the damage to the kernels is not visible.

In an endeavour to obtain the best of both worlds, the manufacturers of some self-propelled combines have increased the widths of their cylinders to somewhere between the earlier narrow cylinder and the full width cylinder arrangements (Fig 220c). One unique combine layout which proved most successful is also shown in Fig 220, d and e being front elevation and plan respectively. In this case the cylinder is arranged longitudinally in the machine so that the crop is fed into the cylinder with its straw parallel to the beater bars. This reduces the power requirement and straw damage, and simplifies grain separation.

GRAIN DISCHARGE SYSTEM

Methods of grain discharge have been outlined previously, but their relative merits have not been discussed. Machines fitted with large-capacity tanks—generally referred to as *tanker-type* combines—are efficient and labour saving where the system of transport and handling is well organised. There must be sufficient tractors and grain-tight box trailers to shuttle between the combine and the grain drying and storage installation. Such a system has, in the past, been considered only justifiable where large acreages of cereals are involved, but many smaller farmers are now adopting bulk grain handling. The tank can be unloaded into an accompanying trailer with the outfit in motion, and there is no need for a second man riding on the combine.

The so-called *bagger-type* machine is still quite widely used, especially by farmers who practise *in-sack* grain drying. The bags are filled and tied on the platform of the combine, and then discharged down a chute in groups to simplify later collection.

Preparation for work

The routine maintenance requirements of the combine during its period of use are discussed later. It is useful, however, at this point, to itemise those features which normally require attention on a machine at the beginning of a new season, assuming that it received proper attention at the end of the previous season.

First, any rust preventive should be removed from components which convey or engage with the crop. Chains and belts removed for storage purposes should be fitted and correctly tensioned. If the machine is first to be used in a clean-strawed crop, a serrated-sectioned knife should be fitted and a spare one should also be at hand. Where under-sown ley or green trash are present in the bottom of the crop, a well-sharpened plain-sectioned knife is often preferable.

The various slip clutches on the machine should be checked to ensure that their elements are not seized. Where p.t.o.-driven combines are concerned, the tractor must be of adequate capacity and have suitable gear ratios. In attaching the tractor, the recommended drawbar dimensions and power-drive shaft support locations should be used. When the above work has been carried out, all guards should be secured in place, and the whole machine checked to ensure that none of its components is obstructed in any way. Even when all the components are apparently clear, the mechanism of the machine should be put into motion with gradual engagement of the clutch or driving belt and run light for a short time before proceeding to the field.

Operational adjustments

Reference has been made to the fact that many of the combine's components are adjustable to accommodate various types of crops and working conditions. On self-propelled machines, some of these adjustments can be remotely controlled from the driver's seat by mechanical levers or hydraulics. Other settings, which

331

need less frequent alteration, are made on or near the respective component or assembly. On trailed-type machines, some operational settings can be made from the tractor driver's seat, but the majority of adjustments have to be made on the machine itself.

There is considerable variation in the precise forms the adjustments take, but their principles and effects are common to most harvesters.

CUTTING HEIGHT

Cutting height is varied by raising or lowering the cutter bar, which is an integral part of the header assembly. Considerable effort is required to increase the cutting height, especially on self-propelled machines which have the additional weight of an auger and steel feed-conveyor. For this reason, most current self-propelled combines employ a hydraulic lift system, while on some older machines, the lift mechanism is powered by an electric motor. Hydraulic lifts are also used on some trailed harvesters while others may have power-assisted or manual lifts.

Whatever type of lift system is used, the weight of the header is always counter-balanced by a powerful spring to reduce the required effort. As a rule, this should be so adjusted that the header requires as much effort to lower it as is required to raise it. It is possible, with some combines, to 'float' the header, in which case the latter follows the ground contours independently of the main structure of the machine. This is a valuable feature when close cutting is required on an undulating surface.

Cutting-height adjustment is usually variable from about 2 in to 30 in (Fig 221).

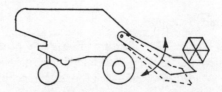

221 Cutting height adjustment—header moves on rear pivot

REEL

Position

The setting of the reel relative to the cutter bar both vertically and longitudinally, is most important (Fig 222). The optimum

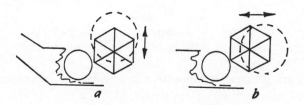

222 Reel adjustment: *a* Vertical, *b* Forward and back

position is determined by two factors: the type of header (auger or straight-through elevator), and the nature of the crop (height, ripeness, density and presence of weeds, etc.).

On auger-type headers, the reel is generally smaller in diameter than on straight-through type headers, and the aim in setting the reel is to feed the crop in to the underside of the auger *butts* first. In normal conditions, the reel setting should be well forward and low in relation to the cutter bar (see Fig 223a). On the alternative type of header, the endeavour is to lay the crop back on to the feed conveyor so that it is fed, ears first, into the cylinder (Fig 223b). To achieve this, the reel must usually be set so that its

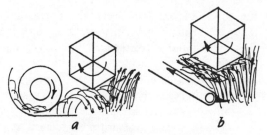

223 Reel setting techniques: *a* Auger header, *b* Elevator header

333

centre is approximately over the cutter bar and at a height which enables the slats to engage on the straw just below the lowest ears. The actual method of positioning the reel varies on different machines. On older machines, it may involve the use of spanners, while on most current machines, either a manual screw crank or a hydraulic control is provided.

Speed

For maximum cutting and intake capacity, the peripheral speed of the reel should be slightly greater than the forward speed of the machine.

In spite of the importance of correct reel speed, some combines are not provided with facilities for its quick adjustment. Some-times, it involves fitting alternative sprockets or pulleys. On other machines, however, the adjustment is simplified by the provision of stepped or adjustable-width vee pulleys. Needless to say, if the reel speed is too low, its effect will be to push the crop away from the cutter bar rather than to gather it in. On the other hand, an excessively high reel speed—particularly when crop conditions dictate a low ground speed—may result in grain being knocked out of the standing crop by the reel slats or tines.

One method used to ensure a constant reel-speed/ground-speed ratio employs a land-wheel drive to the reel, but this arrangement has proved in some cases to be inadequate when working at low ground speeds in laid crops.

Angle of tines (pick-up reel)

Although the so-called standard reel, which has plain wooden slats (Fig 224b) is adequate in clean upstanding crops, the alternative *pick-up* reel, which has tined slats (Fig 224a) is a great asset in laid corn. It is commonly kept on the combine permanently, though its use is not desirable in a very ripe, standing crop on the point of shedding, where its many tines can increase pre-cutting loss of grain from the ear. Its moving parts require lubrication, which is not necessary on the standard reel. A crank at the end of each bar is connected to a circular frame, which, by virtue of its eccentric disposition relative to the reel spindle, keeps the tines in their pre-set position continually, so that clawing or feathering is possible. The precise angle of the tines can be pre-set. When working in a laid and tangled crop, they should be set with

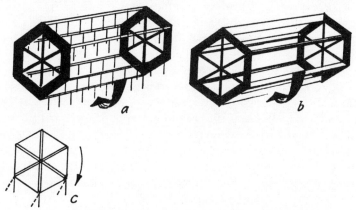

224 *a* Pick-up reel, *b* Standard reel, *c* Rake on tines

backward rake as shown in Fig 224c. When cutting such crops as linseed or seed clover, it sometimes helps to set the tines vertically, or with a slight forward inclination, to perform a sweeping action over the cutter bar.

FEED-TABLE AUGER

The auger—where fitted—always has a small amount of vertical adjustment (Fig 225a). The maximum clearance possible between

225 Auger adjustments: *a* Vertical, *b* Forward and back

M

the periphery of its spiral vanes and the feed tables behind the cutter bar rarely exceeds 2 in. In general, the auger should be raised for cutting long- and coarse-strawed crops. In damp cutting conditions, also, it may be an advantage to raise the auger slightly, particularly when forced to combine crops with unripe straw. Beans and peas need maximum clearance between the auger and the table to avoid excessive shelling of the pods on the feed table. Clearance should be reduced when cutting short and soft-strawed crops, and for combining clover and similar fine seed crops.

On some combines, the auger can be adjusted horizontally, that is backwards and forwards relative to the cutter bar (Fig 225b). In its forward position, it gains quicker purchase on the crop, whilst in its rearward position, the straw is deterred from *wrapping* by the proximity of the spiral vanes to the deflector or cut-off plate. The ideal setting depends upon the volume and stiffness of the straw, and can best be found by trial and modification.

CYLINDER AND CONCAVE

The settings of the cylinder and concave are critical, and so it is advisable to review their functions. These are: the removal of the grain from the ear without bruising or cracking it, and the discharge of as much grain as possible through the concave as the first stage of the separation process.

The available adjustments are the cylinder/concave clearance, the cylinder speed, and the eccentricity of the concave to the cylinder.

The variable factors which dictate the best concave position and cylinder speed are the size of grain or seed kernel, the moisture content of the grain, the degree of maturity of the grain, and the volume and toughness of the straw or haulm.

Cylinder/concave clearance

Detail variations of cylinder and concave design result in individual recommendations regarding this setting. Such recommendations should be regarded only as basic settings, to be modified according to results. The alternative methods of achieving the setting are by raising or lowering the concave in relation to the cylinder, or by raising and lowering the cylinder complete with its bearings in relation to the concave. The first method is favoured by most manufacturers.

Adjustment, in many cases, involves the use of spanners, although more attention is being given to the simplification and speeding up of this setting on modern machines. On some combines, the adjustment can be carried out by remote control from the driving position. Fig 226 illustrates two different settings. It

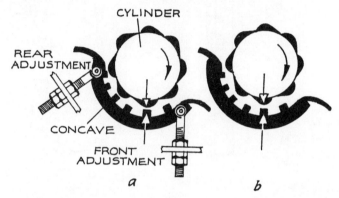

226 Cylinder/concave clearance. Note the wider clearance at the front end

will be noticed that the clearance between the beater bars of the cylinder and the concave bars is greater at the entry end than at the discharge end. This is the eccentricity referred to above; it has the effect of encouraging entry of the crop, while discouraging threshed grain from passing out with the straw at the rear of the concave.

When adjusting the clearance without the aid of an external indicator scale, it is necessary to sight through inspection holes or through a hatch in the cylinder housing.

Cylinder speed

The speed of the cylinder also has an important influence on

337

threshing intensity. Recommendations are specified in terms of rev/min (rpm), but the significant factor is the peripheral speed of the cylinder, i.e. the speed at which the beater bars travel past the static bars of the concave. On some combines, the cylinder speed can be varied continuously through the medium of variable width vee pulleys (Fig 227). This is a valuable facility, enabling

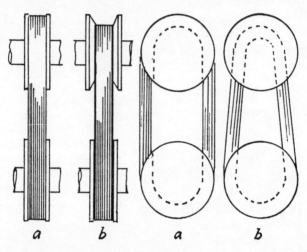

a b a b

227 Speed change by means of variable-width vee-pulleys. Width of pulley determines its effective diameter

the operator to control the threshing severity with considerable precision. On other machines, speed change is by interchangeable sprockets or by stepped vee pulleys. These do not give the same degree of flexibility, and fine-setting of the threshing severity has to be regulated by adjustment of the concave clearance which in some cases is rather time-consuming.

Usually, for threshing cereals, the peripheral speed of the cylinder must be about 6,000 ft per minute. This can be converted into revolutions per minute by dividing the ft/min by the circumference of the cylinder (in feet). From this it will be appreciated that the diversity of cylinder speed recommendations given by different combine manufacturers for any one type of cereal is due to the different cylinder diameters they use. The importance of adhering

338

to the maker's recommendations cannot, therefore, be over-stressed.

Some machines are fitted with a tachometer to record the cylinder speed, but if this is not provided, an inexpensive revolution counter may be used for checking and adjusting cylinder speed.

Concave clearance and cylinder-speed combination

Insufficient concave clearance and/or excessive cylinder speed will result in *over-threshing*, and this may produce one or more of the following effects: cracked, skinned or bruised kernels; an excessive quantity of chaff; excessively lacerated straw.

Excessive concave clearance and/or too low a cylinder speed results in *under-threshing*—all the grain is not extracted from the ear. Too low a cylinder speed may also cause straw to wrap round the cylinder, particularly if the crop is damp, or unripe, or contains an excessive amount of green trash.

Ideally, to avoid unnecessary grain damage, the aim should be clean threshing, with the lowest possible cylinder speed.

Generally speaking, the smaller and finer the grain or seed to be threshed, the smaller the required concave clearance, and the higher the cylinder speed. Conversely, for large easily threshed crops such as peas and beans, a wide concave clearance is necessary together with a reduced cylinder speed.

Further reference is made to these combinations under *Field procedure*.

When combining very fine seeds such as clover, grasses, and linseed, a blanking plate must be fitted to the periphery of the concave in order to prevent the heads or pods from escaping through its grid-like structure and missing the threshing process Fig 215 shows a portion of such a plate. On some combines various proportions of the concave may be blanked off in this way, depending on the nature of the crop, as excessive restriction may result in loss of seed with the straw.

STRAW SHAKERS

The function of straw shakers has been described. Adjustments are

339

not common on these, but provision is sometimes made for the adjustment, addition or removal of triangular fillets (sometimes known as *risers*) on the upper surface of the shakers; and these increase the amount of lift and agitation the straw receives, as it is being conveyed to the rear. Although maximum agitation is always desirable, when soft and damp straw tends to hang on the risers, they should be removed.

Sometimes the slatted surface of the shakers becomes blocked, especially by awns when harvesting barley, and although no adjustments are possible to correct this, it is important to check occasionally, and clean the shakers if necessary. It is also important that the shakers run at their correct speed. The tension of the driving belt should therefore be checked frequently.

TOP SIEVE, BOTTOM SIEVE AND FAN

The functions of the two sieves and fan are so inter-related that an explanation of setting technique must keep all three in focus (see Fig 217). The segregation of the grain from the short straw, chaff, dust and any other trash which reaches the shaker shoe from the concave and grain-return pan, is based on the contrasting sizes and weights of these materials. When correctly set, the sieves screen the larger and heavier particles of trash from the grain or seed, while the fan winnows away the lighter trash. The openings of the top sieve are usually adjustable (see Fig 228), and should be

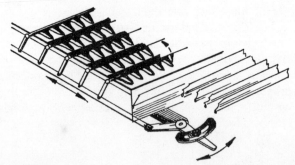

228 Adjustable sieve—the control lever operates all sieve louvres simultaneously

set to allow all the grain to pass through to the bottom sieve but with the minimum of trash.

The bottom sieve may be adjustable like the top sieve or instead constructed of perforated steel and interchangeable. In either case its apertures should be finer than those of the top sieve to intensify the screening process. Although the adjustable type of bottom sieve is more convenient for altering the setting, the plain inter-changeable perforated type is usually superior in performance, particularly in dirty harvesting conditions.

The fan blows a constant controlled stream of air through, under and over the sieves to remove chaff and dust. Both the volume and the direction of the air stream are adjustable (Fig 217).

This sifting and winnowing process is simple in principle, but it is sometimes difficult to get the best combination of sieve and fan settings to suit a particular crop. Consequently, the shaker shoe often becomes the bottleneck in the whole combining process, and may well be the cause of serious grain losses.

Operational guidance is always given by the manufacturer, although precise setting can only be determined by experiment. Small seeds obviously need fine sieve settings, while such crops as peas and beans need wide settings, and the range of cereals which fall between these extremes of size and weight need various intermediate settings.

As far as fan settings are concerned, fine and light seed or grain require a delicate air stream while the heavy haulm of the bean crop usually needs maximum air blast to produce a clean sample.

The direction of the air blast from the fan is controllable by means of one or more deflector valves in the fan outlet ducting. In heavy or dirty cereal crops, best results are usually obtained by directing the air blast fairly steeply towards the forward end of the sieves (the end nearest the cylinder). This enables full advantage to be taken of the length of the sieves, by eliminating most of the lighter material at an early stage. On some machines, the position of the fan in relation to the concave is such that a forward setting of the air stream winnows the grain while it is still in suspension—falling from the concave. In other cases, the shaker shoe is stepped at its forward end so that the grain has to drop through the air stream before it reaches the top sieve (Fig 217). The volume of air delivered by the fan may actually be adjusted in one of three ways: by variable speed rotor, by outlet valves or by intake valves (Fig 229).

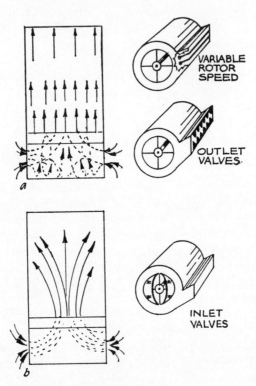

229   Types of fan control and their effects: *a* Control by rotor speed or outlet valves gives more even flow across sieves, *b* Control by outlet valves concentrates air through centre

With the intake valve method, when making modifications to the setting, both valves must be moved so that they are always concentric with the rotor. Operators tend to make the mistake of adjusting one valve only—the one which is most accessible or which moves more easily (Fig 230). This practice results in a distorted air stream through the shaker shoe. Some manufacturers prevent this occurring by coupling the valves to ensure simultaneous movement.

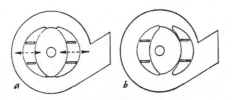

230 Adjustment of dressing fan intake valves: *a* Correct (concentric), *b* Incorrect

Regulation of the volume by variable rotor speed or by outlet valves causes a pressure to build up in the fan housing, and this results in a more even air stream across the width of the shaker shoe as indicated by diagram *a* in Fig 229. End valves tend to encourage a concentration of blast through the centre of the shaker shoe as shown in *b*.

The functions of the shaker shoe are so important that the effects of maladjustments must be explained.

Too coarse a sieve setting will allow too much trash to pass through to the bottom sieve and grain hopper, resulting in blockage and a dirty sample (Fig 231a).

On the other hand too fine a top- or bottom-sieve setting may result in too much threshed grain entering the returns system, and possibly choking it. It may also cause grain to be lost off the rear of the top-sieve extension, especially in heavy yielding crops or when working up a gradient (Fig 231b).

Insufficient volume or incorrect direction of the air blast from the fan may also allow light trash to accumulate on the sieves. This will cause restriction of the apertures and loss of grain off the top sieve, large quantities of trash in the sample, and overloading of the returns system. Fig 231c shows a correct combination.

It is fair to say that more blockages and grain wastage are

343

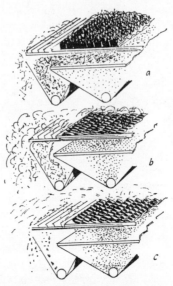

231 Effect of incorrect sieve settings: *a* Top sieve too coarse,
allowing trash to block bottom sieve, *b* Top sieve too fine causing
loss of grain and overloading of returns system, *c* Correct

caused by an inadequate air blast than by an excessive blast, and it
is often the conscientious operator who runs into difficulties,
trying to produce a clean sample in the field. In view of the
urgency of the harvesting operation, the prime aim should be to
eliminate as much restriction to the flow of grain as possible. It is
best to use relatively coarse sieve openings in conjunction with the
maximum safe air blast, and some trash in the sample might even
be acceptable in certain circumstances rather than slowing down
the whole operation or wasting grain. This is not, however, to
condone careless combine setting, but rather to make as the
priority the elimination of all avoidable restrictions.

The efficiency of the combine's separating and dressing
mechanism can be seriously reduced by such factors as damp
harvesting conditions, excessive green trash in the crop and
abnormally brittle straw due to the effects of weathering.

Two additional facilities provided on some shaker shoes, and so far not mentioned, are an adjustable tailboard which, despite its name, takes the form of a steel vertical extension at the rear of the shoe and beneath the top-sieve extension (Fig 217), and a longitudinal attitude adjustment for the top sieve or the entire shaker shoe. Both these adjustments reduce losses of grain when combining uphill, and also reduce the loss of beans or peas which tend to roll along the top sieve surface and out of the machine.

RETURNS SYSTEM

The objects of the returns system have been briefly outlined, but it is important that its functions are clearly understood. These are to collect any ears of grain which escape through the concave unthreshed, and to return these to the cylinder for threshing; and to collect any threshed grain which, as a result of incorrect sieve setting or unavoidable sieve blockage, fails to pass through to the normal receiving hopper.

The system, therefore, may have to handle either or both un-threshed ears and threshed grain. The only adjustments provided, and even these are not found on all machines, are the spacing of the tines or louvres forming the chaffer extension, and the control of the returns outlet from the system. Most combines have an inspection hatch or window at the top of the returns elevator, so that the operator can check on the nature of the returns being conveyed. This is a valuable guide to what is happening in the shaker shoe, and indeed at the cylinder and concave. An alternative arrangement, which helps supervision, has the elevator outlet in front of the cylinder just above the feed conveyor, in constant view of the operator.

Following are some of the materials that may be returned through the system, together with explanations of the possible maladjustments they indicate:

1. Excessive quantity of whole or portions of unthreshed ears. This could be due to insufficient concave clearance at the front, or entry end, or possibly the breakage or loss of one or more of the wire separators which form the concave grid structure.

2. Excessive quantity of threshed grain. This generally indicates inadequate openings of the top and/or bottom sieve(s). Alterna-

tively, the trouble may be due to inevitable sieve choking because of adverse combining conditions. The only remedy here is to slow down the ground speed.

3. Excessive quantity of light trash, such as short straw, caving, etc. This would indicate insufficient or badly directed air blast from the fan.

4. Heavy trash, such as green material, thistle heads, etc. This would suggest too wide a setting of the top sieve extension, and possibly too low a cutting height, causing too many weeds or undersown clover to be taken in with the crop.

These comments show the value of making frequent checks on the nature of the returns, and there is scope for improvement on some current machines in the design and location of visual check points on the returns system.

The outlet from the returns elevator sometimes goes direct to the cylinder, so that whatever is returned has to be rethreshed. On some harvesters, however, the discharge can be made either into the cylinder and concave, or on to the straw shakers or shaker shoe to avoid a second threshing. In bad cutting conditions, such as in a laid and tangled crop, particularly during a wet harvest season and when the intake of some weeds is inevitable, some threshed grain will enter the returns system. If this is returned to the cylinder, considerable grain damage will occur. So, the alternative returns outlet which by-passes the cylinder is clearly an advantage in such circumstances.

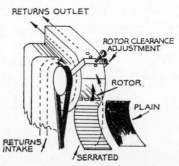

232 Rethresher unit. Intensity of the processing is adjustable to suit the nature of the returns

A revolutionary approach by one manufacturer to this problem of returns treatment is the provision of an independent rethresher unit. The unit comprises a cylindrical housing, inside which is a rotor whose clearance from the periphery of the housing can be varied. The returns are fed into the casing, and the severity of its treatment can be regulated by exposing the returns to either a plain or serrated segment of the cylindrical housing (Fig 232).

GRADING SCREEN

A grading screen is available for most 'bagger'-type machines. Two types of screen are in common use (Fig 218): (a) takes the form of a revolving cylindrical screen. The grain is fed into one end of the cylinder and augered along inside it towards the other end. The spacing between the coils, which form the periphery of the cylinder, can be regulated to allow small seeds and immature kernels to drop out of the cylinder as the grain passes through to the *best quality* section. (b) shows the other type of grading screen. The principle is that of a stationary perforated cylindrical screen through which the grain is conveyed by an inner spiral impeller. Small seeds and immature kernels are discharged from the cylinder by centrifugal force. Intensity of screening in this case is regulated by the use of alternative outer cylindrical sieves having different-sized perforations.

Grading screens are only advantageous in clean, dry harvesting conditions, as moisture from damp or green material contaminates the grain and causes blockage of the screen. In damp combining conditions, the screen should be by-passed, if this is possible. One type of by-pass valve is shown in Fig 218a.

CONTROLS

Reference has been made from time to time in this chapter, to an increasing number of operational settings on the combine being regulated by control levers, as distinct from adjustments involving the use of spanners. Apart from the time-saving virtue of this trend, higher standards of work result, as cumbersome and

347

awkwardly located adjustments discourage the accurate setting which is desirable on such a machine as the combine harvester.

With trailed combines, centralisation of the various controls is more difficult than with self-propelled machines, but it is generally possible for the operator to regulate both cutting height and reel position from the tractor seat by remote-control levers.

On self-propelled combines, in addition to the driving controls which have much in common with other traction vehicles, some or all of the following controls are located within reach of the operator from his driving position: ground speed (often ranging between $\frac{1}{2}$ and 12 miles an hour—the higher end of the scale being used for transportation), cutting height, reel speed and position, cylinder speed, concave clearance, threshing mechanism drive clutch, and grain discharge (tanker models).

A number of the above settings are achieved on some machines by hydraulics.

Power steering and independent brakes are further recent developments which are proving to be most valuable in improving manœuvrability.

Special attachments

The versatility of a combine depends largely on special attachments which are normally available as extras to the basic machine. Most of these are more or less standardised in their form and principles of operation.

PICK-UP REEL

As described previously, the pick-up reel is designed to replace the standard reel on the header when cutting crops which require lifting in front of the cutter bar (Fig 224a). It may be used for harvesting laid and tangled cereals, seed crops (clover, linseed, etc.), and, in conjunction with a pick-up attachment, for bulky windrowed crops such as peas, beans, etc.

The pick-up reel also helps the recovery of the flattened corn when cutting the back-swath with a trailed harvester.

PICK-UP ATTACHMENT

The pick-up attachment must be distinguished from the pick-up reel described above. The function of the pick-up attachment is quite different from that of the reel. Fig 233 illustrates a commonly

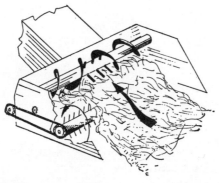

233 Action of pick-up attachment

used type of pick-up attachment fitted to the header in its working position. Its function is to pick up a previously cut crop from the windrow, as described under *Field procedure*. The attachment is usually chain driven from an auxiliary shaft on the combine. The tines work with a retracting or feathering action by means of cams or eccentrics. When the attachment is fitted, the header balance spring(s) should be adjusted to counter its weight. Sometimes the fitting of ballast weights to the rear wheels of self-propelled combines is also recommended, as the rear-wheel steering can be affected if the machine becomes front-heavy by the additional weight of this attachment.

CONCAVE BLANKING PLATE

Reference has been made to the concave blanking plate. It is a curved plate designed to fit snugly around the outside of the concave, to blank off a portion of its grid-like frame (Fig 215). Its purpose is to prevent the heads of such crops as clover, grasses and linseed from escaping the rubbing action of the

349

cylinder. Unfortunately, on most machines, the concave is difficult to get at, and this attachment is often ignored when it could substantially improve threshing efficiency. On the other hand, if a fitted plate is not removed, serious grain losses will occur when the machine next combines a cereal crop.

SPECIAL SIEVES AND SCREENS

Combines are generally supplied with all the sieves and/or screens necessary for the more usual range of farm crops, but special sieves may be necessary for crops which are unusual in that they have ultra-fine seeds or need stringent screening.

STRAW-BALING ATTACHMENT

Several press-type low-density baling attachments have been marketed for attachment to the rear of the combine to simplify the collection and handling of the straw. It is a miniature version of the low-density baler described in Chapter 9. The advantages of such a unit are fairly obvious, but it also has certain drawbacks. These are: (a) extra power demand on the combine or tractor engine—in some cases, sufficient to make the cylinder speed fluctuate on the combine, (b) the possibility of blockage or tying faults holding up combining operations. (These can be particularly frustrating when conditions are otherwise suitable for a good rate of work), (c) on some machines, the position of the attachment obstructs access to the sieves, (d) the additional weight on the rear of the combine may make the steering of a self-propelled machine heavy.

In spite of these objections, however, the attachment may well justify its cost, even for occasional use.

Field procedure

CONDITION OF CROP

The ability to determine the fitness of a crop for combine harvesting is essential if grain damage and loss are to be avoided in the process, and deterioration in subsequent storage. The two main factors to be considered are maturity and moisture content.

Both immaturity and high moisture content make clean threshing difficult to achieve besides increasing the risk of damage to the grain. Even where such a crop is adequately dried, it can be of poor quality due to a high proportion of withered kernels. Over-maturity, on the other hand, results in high cutter bar losses, i.e. loss of grain due to shattering from the ear as the standing corn is struck with the reel and cutter bar. Also, the grain may crack during threshing due to its brittleness.

The moisture content should ideally be in the region of 17% to 21%, especially for a potential seed or malting sample where the percentage and strength of germination are the criteria. It is not always appreciated that grain can suffer bruising damage which may not be apparent on inspection of the sample, but which may have serious effects on the germinative energy and storage qualities of the grain.

Moisture content can be accurately measured only with a meter designed for the purpose. One electrically operated moisture meter records the moisture content by measuring the resistance the grain offers to an electric current passed through it. Other types of meter are available, and most are very simple to use.

To those with wide experience of combine harvesting some of the above remarks may sound rather pious when one considers the laid, tangled, sodden and weed-infested crops which sometimes have to be tackled. Needless to say, in such adverse conditions most of the ideals of combine management and setting go by the board—the order of the day being 'get what you can!' On the other hand, there is no excuse for neglecting good operating technique when conditions are favourable.

COMBINE CAPACITY

Combine capacity obviously has an important bearing on the ability to cut at the right time. Some authorities suggest that a reasonable basis on which to calculate combine requirements is 30 acres per foot of cutter-bar width. In some seasons and in some parts of the country, however, even this figure may be too high, and reducing the ratio may not be as expensive as it may at first appear. Ample combining capacity means that full advantage can be taken of the best weather during the harvesting season. Also, the policy of cutting only during the best hours of the day can substantially reduce drying costs.

Whatever the combine capacity on a particular farm, good management, in terms of forward planning and the careful selection of seed varieties which give some staggering of maturity and phasing of harvesting, can ease the problem of accurate timing of the cutting operations.

When several crops become ready for harvesting simultaneously, it may be difficult to decide which should take priority. Provided it has no tendency to lodge, wheat will usually stand delay better than barley or oats. The weaker strawed varieties of barley tend to *neck*, that is the straw collapses at a point which allows the ears to droop below normal cutting height. Some varieties of oats shed very freely when over-ripe, resulting in heavy field and cutter bar losses.

COMBINING FROM WINDROW

In one method of harvesting, the crop is precut either with a special windrowing machine or with an adapted binder, and left lying in a continuous windrow for the combine to pick up and thresh in a subsequent operation. This procedure is advantageous in the following circumstances: (a) when the crop to be harvested ripens unevenly, as in the growing of mixed cereals, (b) for harvesting clover and grass seed crops where the stems and foliage should wilt before threshing, (c) for harvesting cereal crops in which undersown ley has developed strongly (in this case, it is again desirable to wilt the green material before the crop is threshed), (d) when retrieving a badly laid crop, where the cereal is long and has lodged in patches.

In the pre-cutting operation, a long stubble should be left wherever possible to keep the windrow clear of the ground and allow the air to circulate freely under and through the windrowed crop.

Except in very settled weather, the advance windrowing of large acreages is unwise, as once the crop is cut and in windrow, it takes very much longer to dry out than when it is standing.

To harvest the windrows, the front of the combine has to be modified by fitting a pick-up attachment described previously. The width of the attachment varies on different machines, and it is desirable when forming the windrows, to make their width match that of the pick-up. This is especially important with wide-cylinder combines, where a narrow concentration of material through the centre of the cylinder and concave may tend to spring the latter, resulting in poor threshing efficiency and possibly damage to the concave. To avoid this, at least one manufacturer has introduced a reciprocating finger at the top of the feed conveyor to keep the material spread over the full width of the cylinder and concave.

The ideal height for the pick-up attachment is as high as clean picking up of the crop will permit. This avoids undue strain on the tines and the picking up of stones.

OPENING OUT THE FIELD

Self-propelled machines

Opening out with a self-propelled combine seldom presents any problems. The machine is simply driven round the perimeter of the field in the most convenient direction, depending on the location of the grain tank or bagging platform. If the projection of either of these units is likely to foul the hedge or fence, the machine should be driven round in the direction opposite to that normally followed in the routine cutting. Where high hedges delay the drying of the crop adjacent to them, it might be desirable to make the first circuit a few yards in from the boundary, leaving the circuit under the hedge for a few hours to dry.

Trailed machines

Opening out with a trailed machine usually involves a *back-swath*, which means the grain is flattened by the wheels of the outfit on its first circuit round the field. The back-swath is cut with the outfit travelling in the opposite direction to the initial round. By this means, with the cutter bar set low, almost all the flattened grain can be retrieved. The initial cut may be made in either direction around the field. If the cutter bar is offset to the right of the tractor, and the initial circuit is clockwise, the tractor travels close beside the field boundary. This has the advantage that cutting may continue until such time as the back-swath lying immediately against the hedge has dried out.

The alternative method of opening out—making the first cut in an anti-clockwise direction around the field cutting the crop adjacent to the hedge—may sometimes be preferable. When this method is used, the back-swath is cut on the second circuit of the field, in the routine clockwise direction. An advantage of this is that the corn is often cleaner a short distance in from the boundary, and as the cutter bar has to be set very low to pick up the flattened grain, the knife is less likely to choke, and green trash is less likely to be taken in.

CORNERING

Some operators of self-propelled combines negotiate the corners of the crop by making a *loop turn* through 270° (see Fig 234a). Others make a *righthand turn* at each corner of the crop (as shown below in Fig 234b), and bring the cutter bar back into work as quickly as the steering lock will allow (Fig 234b). This method is also commonly used with trailed harvesters. Another method is to stop at each corner, and reverse to line the outfit up with the next side to be cut, virtually a *three-point turn* (Fig 234c). In any event, the combine should not be run idle by keeping it out of the crop for long periods at the corners, as this has two undesirable effects on the work. The reduced grain flow through the constant air blast results in some grain being blown over the sieves and out of the machine, and some adjustable type sieves may admit short lengths of straw through their openings when the normal train of material ceases to flow. This results in trash entering the grain tank and spoiling the sample.

With p.t.o.-driven trailed combines, short turning is restricted

354

by the power-drive shaft; care must be taken not to strain the universal couplings by turning too sharply.

DRIVING TECHNIQUE

In good harvesting conditions, driving the combine or tractor is a simple operation, but, ironically enough, it is in such favourable conditions that most grain is wasted through over-loading. The high intake capacity of modern combines in normal upstanding crop conditions makes it easy to overload the straw shakers and

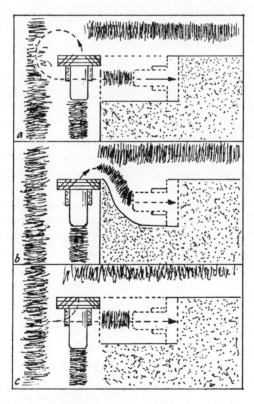

234 Methods of cornering with combine: *a* Loop turn, *b* Righthand turn, *c* Three-point turn

shaker shoe. This is particularly true when combining from windrow.

Such factors as length of straw and density of crop must be taken into account when regulating ground speed. In order to ensure an even intake of the crop, it may be necessary in patchy crops to increase and decrease ground speed from time to time according to the crop density.

It is usual to set the cutter bar just above the level of any green trash or under-sown ley present in the bottom of the crop. However, there may be times when long, coarse straw makes it desirable to increase the cutting height in order to reduce the volume of straw taken into the machine. This, of course, leaves an abnormally long stubble which may hinder the subsequent ploughing operation. Combining badly laid crops requires an exceptionally low cutter bar setting and, in such conditions, the ability of the cutter bar to float on skids—a feature of some modern machines—is a distinct advantage, especially on uneven land.

It is desirable to keep the full width of the cutter bar in work at all times. This is because any portion of the cutter bar running light tends to clip short lengths of straw from the stubble left on the previous bout, and these can be difficult to separate from the grain in the sifting process. Also, on wide-cylinder machines, it is better for the full width of the cylinder to be under load at all times. One difficulty, when cutting across a gradient with a non-auger combine, is that the crop tends to slide across the feed conveyor to the lower side of the machine and to one end of the cylinder.

Combines are designed to cope with undulating land, but steep gradients are difficult to manage without some loss of grain. If the machine travels across the hillside, the grain tends to slide down to one side of the sieves. Although, as mentioned previously, some manufacturers provide ledges along the sieves to check this, it is advisable also to reduce ground speed so that the sieves are not loaded to full capacity. Sometimes, better results are obtained by working up and down the hillside, although adjustment is then usually necessary on the longitudinal attitude of the sieves, the tailboard, or the shaker shoe as a whole—according to the facilities provided.

OPERATIONAL CHECKS

Although the combine can be basically prepared for a particular crop before going to the field, precise adjustments can be made only by experiment in the particular crop being harvested. Frequent checks therefore need to be made when the machine is first put into work, and at regular intervals throughout the day. Such checks should be methodical, and a suitable sequence of examinations follows:

Grain losses with the straw

Grain may be lost with the straw for two reasons: incomplete threshing of the ears, or incomplete separation of loose grain from the straw. So the ears should be closely examined, particularly the base of the ear which is harder to thresh. Also, the ground underneath the windrow of threshed straw should be examined for the presence of loose grain. The presence of grain on the ground may, of course, be due to cutter bar losses, or to shedding from the standing crop, in which case the stubble ahead of the straw discharge point should first be inspected to assess losses with the the straw. Loss of loose grain with the straw may be due to overloading of the machine by excessive ground speed, to partially blocked straw shakers—a particular problem with barley—or to insufficient straw-shaker speed because of a slipping belt.

Losses of grain from the shaker shoe

Any grain losses from the shaker shoe can be checked by a person walking behind the machine in work, and observing the sieves and catching the material being discharged from the shoe either in his hand or in a tray. In clean, dry harvesting conditions, the rearmost quarter of the top sieve should be clear of grain and chaff. If not, the sieve should be opened slightly to pass the grain more freely. If grain is being blown out, the air stream from the dressing fan should be reduced. Conversely, if light chaff is seen to be riding along the top sieve and dribbling off its end instead of blowing clear, this would suggest inadequate air blast or a combination of this and inadequate sieve openings.

Grain sample

The presence of light, dry trash in the sample indicates too wide a sieve setting, and/or insufficient air blast. Damp green trash in the sample also suggests too wide a sieve setting, although this fault could be aggravated by the intake of large quantities of green material resulting from too low a cutter bar setting. Cracked or skinned grain in the sample could be caused by over-threshing due to excessive cylinder speed or insufficient concave clearance, or to incorrect setting of the top and/or bottom sieves—resulting in excessive returns and re-threshing. With a new machine, the sharp edges of grain elevators and augers can also contribute to the damage of grain, especially when the latter is in a susceptible condition.

Nature of returns

The significance of the nature of the returns has been discussed under *Operational adjustments*. Frequent checks should be made on the material being circulated through the returns system, as this reflects the performance of both the threshing and the separating mechanisms of the combine.

Maintenance

Because of the numerous moving parts of the combine and the high speed at which many of them run, the machine's maintenance requirements are rather demanding, although manufacturers are endeavouring to reduce them to a minimum. However, as little combining is done before ten or eleven o'clock in the morning, there is little excuse for neglecting the machine's daily attention. Indeed, an extra half-hour with a grease gun or spanner might well pay a double dividend—first in preventing damage to grain cut in an unfit state by too early a start, and second, by rectifying simple mechanical maladjustments before they give rise to more serious trouble.

Even a machine which has been correctly stored during its idle months will need some attention before starting a new season's work (see *Preparation of the combine harvester for work*).

DAILY ATTENTION

Reference must be made to the maker's instruction manual for details of servicing procedure for any individual machine. However, it is worth outlining some of the items which are common to most machines.

The engine, where there is one, should be replenished as necessary with fuel, engine oil and water; the air cleaner should be serviced and the radiator screen cleaned. The lubrication should be carried out systematically, progressing round the machine and taking care not to omit the less obvious and less accessible points. Chains and belts should be checked for tension and alignment and corrected if necessary.

A brief check should be made each morning for loose nuts or bolts, or undue play on any moving parts—the lack of two or three turns on a reel slat nut has often caused serious damage to a cylinder and concave assembly! Straw shakers and sieves should be examined to ensure that they are completely clear. The trough of the stone trap should also be cleared—even an accumulation of chaff will render it ineffective.

When cutting stops at the end of the day the machine should be run long enough to allow the elevators and augers to clear.

PERIODIC ATTENTION

In the course of the season's work, a number of engine oil and filter changes may have to be made, and once again the importance of adhering strictly to the maker's guidance cannot be over-stressed—bearing in mind that combine engines are subjected to sustained load for long periods in hot and dusty conditions.

Oil levels of the transmission system units and hydraulic

system (where applicable) should also be maintained as specified, and tyre pressures should be checked from time to time.

END-OF-SEASON ATTENTION

When corn harvesting finishes for the season, there are many other urgent jobs following hard behind, and this accounts for the procrastination which proves so costly to farmers in terms of machine deterioration. However, a list of some of the more important items will indicate the importance of end-of-season attention, besides serving as a guide to those with enough determination to make the time to carry them out.

1. As with all machines which handle or process any form of vegetative matter, the combine should be thoroughly cleaned internally and externally, and its crop-friction surfaces coated with a rust preventive. Crop residues left on sheet-metal components can cause serious corrosion during the winter period. Besides causing permanent damage to the structure of the machine, corroded surfaces cause blockages when the machine is next put to work. Those components which can be easily removed from the machine should be withdrawn to simplify the work.

2. All belts should be removed, checked for condition and stored away, and the belt faces of all steel vee pulleys should also be protected from rust.

3. All external chains should be removed, checked for condition and ideally stored in a lubricating oil, and sprockets should also be checked for wear. Internal chains and sprockets, such as the feed conveyor and the grain and returns elevators, should be examined for condition.

4. Before withdrawing the knife from the cutter bar, it is worth checking the knife register to see if it complies with that specified in the instruction book. As described in Chapter 7 the term *knife register* refers to the position of the knife sections in relation to the fingers at the extremity of each stroke. Methods of adjustment vary on different machines, but any discrepancy is usually simple to correct if the maker's instructions are followed. When the knife has been withdrawn, it should be checked for loose or damaged sections. Loose sections can often be detected by a rattle when the

knife is held vertically on its end and given a sharp tap on a hard floor. The knife should be protected against rust, and stored in a protective guard or some other safe place. At this stage, the fingers of the cutter bar should also be checked for condition, alignment and tightness.

5. The cylinder and concave should be carefully inspected for wear and damage. In normal circumstances, the rasp-type beater bars are unlikely to show serious wear as a result of a season's work, but considerable damage may be suffered by the beater bars from stones, and it is not uncommon for these to be picked up when combining from windrow on stony land, despite the provision of a stone trap. Unless a machine is relatively new, it is unwise to replace just one or two damaged beater bars since this is likely to upset the balance of the cylinder. Also, unless the other bars are packed to match, the new ones will stand proud, and upset the concave clearance setting. Needless to say, if stones do enter the threshing assembly, the concave also may be damaged— either the cross bars being bent or the wire separators broken. In such circumstances, the repair of these components is best left to the agricultural engineer.

As well as damage by stones, the concave of a machine can become badly sprung. This is likely to occur when cutting a badly laid crop, when tangled wads of material are taken into the machine. Such uneven feeding of the combine can also arise when harvesting from windrows, especially if the latter lack uniformity.

In any event, even with normal wear and tear, eventually the beater bars will need renewal and the concave will need trueing up or replacement. This work, also, is the province of the agricultural engineer.

On the less common flail-type cylinders, it is the rubber beater-bar facings and concave blocks which suffer damage.

6. A general check should be made over the whole machine, examining all fast-moving bearings, the shaker-shoe eccentrics, straw-shaker bearings, feed-table auger, grain augers, elevator flights, etc.

7. With the machine parked in its storage position, ideally under cover to avoid internal corrosion, the header balance

spring(s) should be relieved of their tension by propping the cutter bar safely in a raised position.

8. Where an engine is fitted, it will require the usual storage procedure, including draining the cooling system, draining the engine oil and replenishing and, where applicable, removing the battery for periodical charging. If the combine has done a few seasons' work, it may be advisable to consider having the engine overhauled.

REPLACEMENT PARTS

However well a combine is designed or maintained, some of its components are more vulnerable to wear and damage than others, especially in a difficult season when the machine is subjected to abnormal working stresses. In view of the costliness of hold-ups, a few so-called 'fast-moving' spares should be kept on hand, even for a new machine. The following items are suggested: two knives, a few knife sections and rivets, two or three fingers, one or two slats for the standard-type reel and a few tines for the pick-up type, a few links to match each of the chains on the machine, slip-clutch elements and springs. For machines not fitted with slip clutches on their grain and returns elevators, spare vee belts for these drives should be stocked.

Some farmers go to the length of holding a spare set of beater bars and a concave. This may be justifiable on very stony land and where more than one machine of a particular model is used on one farm.

Safety precautions

A powered combine harvester has numerous potential danger points, some being fairly obvious, others more subtle. Unfortunately, it is often the enthusiastic operator who gets himself involved in the combine's mechanism, usually as a result of his preoccupation with the job in hand. For this reason, the advice offered here takes into account some of the more common causes of accidents with combines.

1. Ensure that all guards are correctly fitted before work begins, and after any adjustments necessitating their removal.

2. Do not attempt to either lubricate or adjust the machine's mechanism while it is running. This, of course, excludes those adjustments which are made by remote control.

3. Do not smoke when refuelling petrol-engined machines. Clear the exhaust manifold of chaff and straw frequently.

4. Avoid placing limbs under the feed table when carrying out repairs or when fitting the pick-up attachment.

5. Ensure that the threshing mechanism is out of gear, and even then check that no one is in contact with moving parts of the machine before starting the engine or p.t.o. drive.

6. Do not pull away or stop abruptly with a man riding on the bagger platform.

7. Never attempt to clear blockages at the cutter bar, feed table or grain elevators with the mechanism running.

8. Ensure that brakes are efficient and drive with extreme care particularly on hilly land and on the highway.

9. Do not leave the feed table in a raised position when the machine is to be unattended for any length of time, except of course for storage, when it should be supported with stands or blocks.

10. Never carry guns on tractors or combines.

The transition from manual to mechanical potato harvesting is difficult and slow compared with other aspects of farm mechanisation. The greatest problems are the segregation of haulms, stones and clods from the crop, and the removal of soil from the tubers without damaging them. These problems are in part due to the irregular size and shape of the tubers, and the differing types of the soils in which they are grown. However, the range of diggers and harvesters now available provides the farmer with a fairly wide choice from which to select a suitable machine. These machines vary considerably in their structure and working principles, but most of the ones in current production fall into one or other of the following categories: spinner-type digger, elevator-type digger, or complete harvester. There are, however, a few other machines which depart from the conventional pattern.

Construction and operation of spinner-type diggers

CONSTRUCTION

The spinner-type digger is a widely used machine which lifts the tubers from the soil and spreads them on the ground surface for subsequent collection by hand. Its simple design, typified in Fig 235, will work under the most difficult conditions. It consists basically of a wide triangular share, with a tined rotor—generally referred to as the *spinner*—positioned behind it to sweep the tubers to one side. The rotor is driven by the tractor p.t.o. through a slip clutch. Some machines of this type have a second control wheel arranged eccentrically to the spinner to give a feathering action on the tines (Fig 236 and Plate 15). The screen suspended

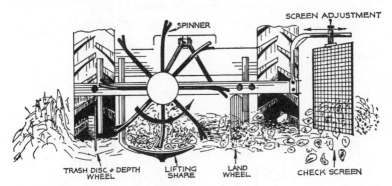

235 Spinner-type potato digger

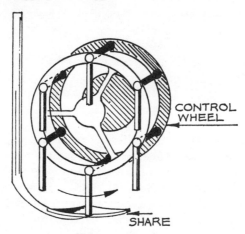

236 Feathering spinner. Tine disposition is maintained by a control wheel eccentric to the spinner wheel

365

on one side of the machine checks the tubers as they are thrown against it, preventing them from being scattered too widely, and at the same time partially cleaning them by the impact. A trash disc is sometimes fitted to cut overhanging haulms in line with the share support arm to prevent choking. Alternatively, a flanged wheel serves the combined purposes of cutting trash and regulating the depth of digging. On some machines, the share support arm is placed behind the rotor to avoid choking.

Some spinner diggers are fitted with a second spinner to discharge the soil and leave the tubers better exposed on the ground (Fig 237).

237 Spinner digger with second rotor to discharge soil and haulms from tubers

OPERATIONAL ADJUSTMENTS

Setting to row width

This is an important setting necessary to ensure that the lifting share travels immediately under the centre of the row of potatoes. On trailed diggers, the wheel on the crop side of the machine should be slid on its axle and secured to run midway between two rows when the lifting share is central with the row being lifted. On mounted diggers, the significant setting is the position of the lifting share in relation to the tractor wheels on the crop side of the row being lifted. Some diggers throw the tubers to the left and others to the right, so the measurement should be taken from

the centre of the share to the appropriate tractor wheel (Fig 238).
Adjustment is made by sliding the digger on its cross-shaft or
lateral beam.

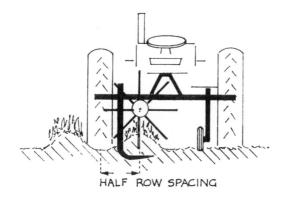

HALF ROW SPACING

238 Digger share centralised on row

Levelling of digger

The level of the mounted-type digger is affected by the adjustment of the right-hand levelling screw of the tractor linkage. For
normal working, the share should be level. Viewed from the
rear, the share stalk should therefore be perpendicular to the
ground.

Depth

The depth should be adjusted so that the share gets right under
the tubers without missing or slicing the lower ones (Fig 239). If
it is too deep, more soil will be lifted with the crop, and this should
be avoided in heavy, sticky conditions, but in light, clean conditions
it can be exploited to cushion the impact of the spinner tines on
the tubers, especially with tender-skinned early crops.

On trailed machines, the depth is regulated by a hand lever,
while on mounted versions, the setting is made either by the
tractor hydraulic control or by depth-control wheels.

N   367

Pitch

Pitch is usually pre-set on trailed diggers, but on mounted machines, variation of the top link length alters the angle of the share (Fig 239). In some conditions, a slight increase of pitch can help separate soil from the potatoes, leaving the latter better exposed for subsequent hand picking.

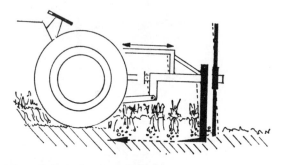

239 Setting of share depth and pitch

Speed of spinner

The speed of the spinner is a most important factor in lifting potatoes. Excessive speed will result in considerable damage to the tubers, due to the severe impact of the tines. Young early potatoes are extremely susceptible to bruising and skinning, especially on light soil where they come out clean. Very low speeds, on the other hand, may fail to leave the tubers sufficiently exposed. On trailed diggers, the spinner is ground-driven, and its speed is governed by the ground speed of the outfit. So ground speeds, with such machines, must be carefully controlled. On some machines, the spinner speed is varied by the changing of sprockets. With power-driven mounted machines, careful selection of the ground-speed/p.t.o.-speed ratio is equally important.

Setting of check screen

The distance of the check screen from the lifting share can be

368

varied to suit working conditions. The best range is that which leaves the tubers well exposed but in a relatively narrow windrow. Very delicate tubers, particularly on stony land, should not be thrown too far.

Construction and operation of elevator-type diggers

CONSTRUCTION

Figs 240 and 241 illustrate a typical two-row elevator type machine. Basically, it consists of a fixed lifting share, similar to that used on the spinner-type digger, and an *open-web elevator*, which is a

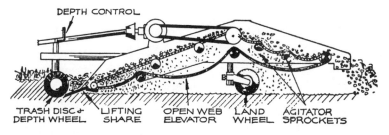

TRASH DISC &   LIFTING   OPEN WEB   LAND   AGITATOR
DEPTH WHEEL   SHARE   ELEVATOR   WHEEL   SPROCKETS

240 Elevator type potato digger

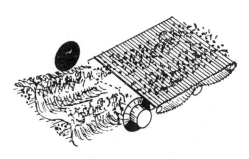

241 Two-row elevator digger: working principles

conveyor made up of rod links (Fig 242). This type of machine is well established in many large-scale potato-growing areas. It lifts the tubers, cleans them quite effectively, and lays them on the surface of the ground in narrow windrows for hand gathering. Both single and double-row versions are available, and their lifting rates can be very high in normal working conditions. On

242 Rod links forming portion of open-web elevator

abrasive soils, however, the rate of wear on the elevator web can be high, and on very stony land repeated blockage of the web can be a problem.

Trailed, semi-mounted and mounted machines are available. Most of them are powered by the tractor p.t.o., and vee-belt drives are commonly superseding chain drives to reduce the high rate of wear suffered by the latter when working in abrasive soil.

Mechanical protective devices

P.t.o.-driven diggers invariably have a slip clutch incorporated in the main drive and most of them also have stone guards at the lower intake end of the elevator to reduce web blockages and damage.

OPERATIONAL ADJUSTMENTS

Setting to row width

The importance of correct lateral setting of the wheels and lifting shares has been discussed for spinner-type diggers. The same factors must be borne in mind when operating an elevator digger.

Depth

Depth is controlled by means of an independently adjustable depth wheel on each side of the lifting share (Figs 240 and 241). On some elevator diggers, the rear wheels also are individually adjustable (Fig 243), and this enables the machine to be set level when working across hillside fields. The optimum share depth

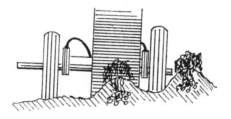

243 Hydraulically-operated independent rear wheel setting

depends upon soil and crop conditions. As with spinner-type diggers, excessive depth should be avoided when operating in heavy and sticky soils, although on lighter land, extra soil can be lifted to protect the tubers from undue abrasion on the elevator. Mounted machines are lifted at the bout ends by the tractor three-point linkage, and some trailed and semi-mounted models have a remote hydraulic ram in the digger to raise the lifting share independently.

Pitch

The facilities for adjusting pitch vary with the type of machine in question. On some mounted and semi-mounted diggers, the top link length can influence the angle of the lifting share. On trailed machines, the share-carrier bracket is usually adjustable.

371

Severity of cleaning

Control of agitation of the tubers may be regulated in several ways: (a) by varying the number or pattern of elliptical idler sprockets carrying the web (Fig 244); (b) by raising and lowering the front

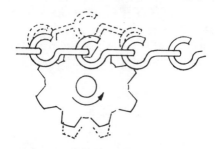

244 Elliptical agitator sprocket

web sprockets immediately behind the share; (c) by dividing the web, and in effect forming two conveyors—the rear one being positioned lower than the front one and so subjecting the tubers to a drop (Fig 245); (d) by varying the elevator speed—this can be

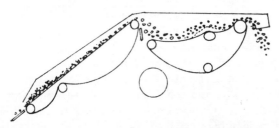

245 Typical arrangement of elevator web when divided for extra cleaning effect

done within limits on p.t.o.-driven diggers by selecting an alternative gear on the tractor. Ground-driven machines may have alternative sprocket or pulley sizes in the drive mechanism. Rubber covered rods are available for the elevators of some

372

machines, for use in a delicate crop where the agitation might otherwise cause excessive abrasion of the tubers.

Potato discharge

On most diggers, adjustable deflector plates help to place the tubers in neat narrow rows on the ground, although some machines are available with a side-discharge conveyor.

Construction and operation of complete potato harvesters

There are several makes of machine which qualify as complete potato harvesters, and the peculiar problems presented by varying soil types and working conditions, and by the inconsistency of tuber size, has led to considerable diversity of design. Variations exist in the digging mechanisms of currently marketed machines and some harvesters require a larger crew than others to keep them working to capacity. The greatest variations, however, exist in the systems employed for cleaning the tubers and for separating them from haulm, trash, stones and clods, and Figs 246 and 247 serve to show the contrasting layouts of two of a number of successful machines. The initial digging and elevation on the first of these is similar to that of the elevator digger already described. When the

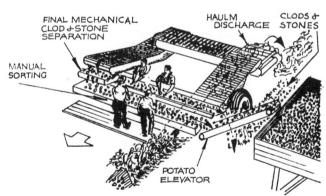

246 Complete potato harvester with elevator digger system and rod-link conveyor separators

373

crop is discharged from the first elevator, it travels along further conveyors as indicated, at one point discharging the haulms and at another segregating the tubers from clods and stones. Operators riding on the machine supervise the final separation and assist manually if need be. The tubers are conveyed into a hopper for bagging off or into an accompanying trailer.

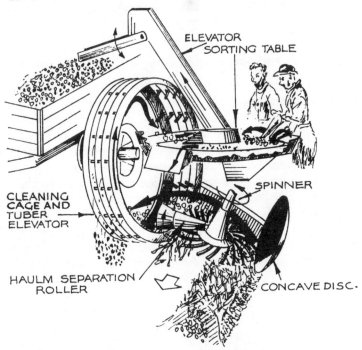

247 Complete potato harvester: with disc, spinner and cage-type elevator/separator

The second machine operates on an entirely different set of principles. The tubers are lifted and deflected by means of a concave disc set slightly oblique to the direction of travel. Simultaneously, a tined wheel rotating in the horizontal plane conveys the tubers into a revolving cage, removing soil in the process. Haulms carried round with the rotor are ingeniously segregated from the tubers by means of a roller running snugly against the face of the tined rotor. The tubers then enter a vertically revolving cleaning cage, which also acts as a conveyor to lift them up on to a flat revolving sorting table.

Operators, riding on a platform, transfer the tubers from the rotating circular table to a trough round the edge. From this trough, the tubers are discharged into a bagging attachment or into an accompanying trailer, by means of an elevator.

In both these machines, as indeed in most other complete harvesters, many rubber or rubber-covered components are used to keep tuber damage down to the minimum.

A detailed description of a particular machine would be of little value, in view of the lack of standardisation in the general layout of harvesters, and in their working principles. However, some of the more common features employed in modern machines are worth mentioning.

Such features include hydraulic lift for digging shares; independent height adjustment for rear land wheels so that the machine can be set level when working across a gradient (Fig 243); alternative types of digging shares or discs to suit working conditions; adaptability of the manual sorting layout so that operators may remove either the tubers from the clods and stones, or the clods and stones from the tubers—whichever is easier; elevator attachments with variable discharge heights to avoid damage from long drops into accompanying trailers; haulm-cutting or pulverising attachments.

Mechanical protective devices

Some form of slip clutch or shear-pin is incorporated in most of the principal drives, and immediate attention to these is most important when they become operative through obstructions. Vee belt drives are also commonly used as safety devices.

OPERATIONAL ADJUSTMENTS

The operational settings of the digging mechanisms of complete harvesters may vary in method, but the basic principles and their effects are similar to those of the spinner and elevator-type diggers

o   375 discussed previously. As outlined in connection with those machines, the digging shares should be set to a depth just below the lowest tubers, although it is important to avoid digging too deeply with the complete harvester where the soil is cloddy, as this can make a considerable amount of extra work for the separating mechanism and the sorters.

With the cleaning mechanism, too, the objectives are as described previously, although the methods may vary considerably from one harvester to another. It may sometimes be necessary to choose between accepting a less clean sample and one with a high percentage of damaged tubers. Needless to say, the former is preferable, although in difficult circumstances it is important that operators have the knowledge and ability to exploit all possible adjustments to the full.

Field procedure

OPENING OUT THE FIELD

As with most harvesting operations, the field system employed for potato lifting depends upon the proposed method of crop collection and transport, the location and type of storage, and the labour available.

The headland rows must be lifted first, before the straight work can proceed in lands. The wheel track of the tractor, and of the digger or harvester, must be set to match the row spacing, and with mounted and semi-mounted machines, it is usually necessary to restrict side-sway on the tractor lower links by tightening the check chains or by fitting stabilisers. As with other p.t.o.-driven machines, the power-drive shaft should be correctly aligned and of the specified length.

OPERATIONAL CHECKS

In the course of the work, frequent operational checks should be made.

Lifting efficiency

If tubers are left in the ground, this is usually due to inadequate depth or incorrect alignment of the lifting share. If the share depth does not respond to the operating lever or depth-wheel adjustment, this may be due to a worn share point, insufficient pitch of the share or excessively hard land.

Condition of tubers

Excessively dirty tubers indicate insufficient agitation; bruised and skinned tubers indicate excessive agitation. However, as mentioned previously, a compromise has sometimes to be made. Methods of regulating the severity of treatment have already been outlined.

Tuber windrows

With spinner diggers, the width of the windrow produced should be controlled to avoid unnecessary scattering of the crop. At the same time, the tubers should be left exposed for the hand pickers —if the windrow is too narrow, they may be covered with soil. The width of windrow is usually regulated by a flexible curtain or by a screen as shown in Fig 235. The speed of the spinner also affects the scattering of the tubers.

Operational efficiency is usually assessed on the basis of (a) man-hours per acre, (b) man-hours per ton, (c) leavings (the proportion of crop left in the ground due to inefficient lifting), and (d) tuber damage.

The term *scuffing* is sometimes used to describe superficial damage to tuber skins of a degree unlikely to cause deterioration in storage.

Maintenance

DAILY ATTENTION

The lubrication points of the spinner digger are easily seen and

377

generally quite accessible. On some elevator diggers and harvesters, sealed bearings are used, and care must be taken not to over-pressurise these. It is advisable not to lubricate elevator chains as the oil attracts abrasive dust.

PERIODIC ATTENTION

Items which should be checked periodically include: chain and/or belt tensions, efficiency of slip clutches, and tyre pressures. On complete harvesters (where applicable), the various conveyors should also be checked for condition and tension.

END-OF-SEASON ATTENTION

At the end of the season, the machine should be completely cleaned and inspected, paying particular attention to all moving parts. Before storing away, all the bearings should be greased, and the whole machine protected against corrosion.

REPLACEMENT PARTS

With the simpler spinner or digger machines, spare share points and chain links are usually all that is required, but with many of the more involved complete harvesters, it is advisable to be guided by the manufacturer as to which parts to keep on hand. Some makers offer a list, or emergency pack, of the items more susceptible to wear and damage.

Safety precautions

The principles and operation of potato-lifting machinery are, in a number of respects, similar to those of sugar beet harvesters, and most of the comments made on the subject of safety in connection with those machines (see Chapter 12) apply equally to potato diggers and harvesters.

In recent years much research has gone into the development **of** sugar beet harvesting techniques, and a good deal of the credit **for** this work must go to the British Sugar Corporation, which **has** played an extremely active part in the organisation of demonstrations on a local and national basis. The objective at such events has been to ascertain the best way to top, lift, clean and collect the beet with the minimum of hand-labour, of damage and loss of beet, and of dirt and top *tares*—in the face of the varied soil conditions and adverse weather experienced during the beet-harvesting period of October to February. The land can be anything from dry and hard to wet and soft, severely frosted or even snow-covered.

A small proportion of the sugar beet crop is still harvested by hand. A somewhat larger acreage is gathered by a combination of hand and machine operations—tractor-drawn beet ploughs, sledges, or spinners being used to loosen the beet or lift them to the surface of the ground. In another method, usually referred to as the *two-stage system*, two machines are used—the first to top the beet in the ground, and the second to lift the beet. The majority of the crop in this country, however, is harvested today with the so-called *complete harvester*, and this is the machine that is considered primarily in this chapter. (See Plate 16.)

Fig 248 shows a typical machine, embodying a topper unit, a top-clearing device (sometimes replaced by a top-collecting attachment), root-lifting components, elevating and cleaning

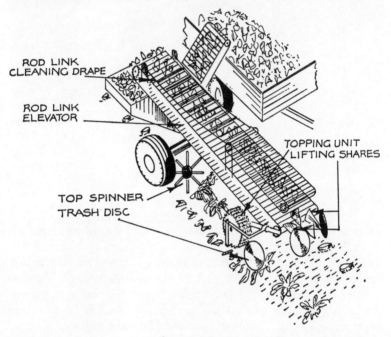

ROD LINK
CLEANING DRAPE

ROD LINK
ELEVATOR

TOPPING UNIT
LIFTING SHARES

TOP SPINNER
TRASH DISC

248 Layout of typical complete sugar beet harvester

mechanism, and a hopper to receive the beet, or an elevator for periodic or continuous discharge into an accompanying trailer.

The working principles of these various assemblies can be better appreciated when their functions are understood, and an explanation of these follows.

Construction and working principles of the complete harvester

TOPPING UNIT

The function of the topping unit is to remove the crowns of the beet together with their leaves, buds and shoots, but without too much of the valuable sugar-producing root. Certain variables have to be accommodated in this operation, and these are the irregularity of the size of the beet, the unevenness of the height of the beet, and the condition of the soil which supports them. Also, the density and coarseness of the tops can considerably influence the performance of the unit. Various adjustments are possible and these are dealt with later.

Fig 249 illustrates the action of the topping unit. A *feeler wheel*, comprising a number of wheel-like elements with serrated edges, rides along the top of the row of beet. It is free to rise and fall with the varying heights of the beet, as the machine progresses. Attached to the feeler-wheel frame is the knife which slices off the tops. So, as the wheel rises and falls, it automatically

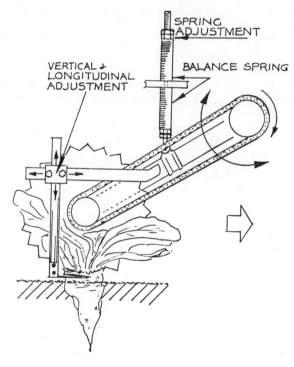

249 Action of topping unit

gauges the cutting height of the knife on each individual beet. The feeler wheel is driven by one of the land wheels of the machine so that, as the serrated edges of its elements grip the crowns of the beet, the latter are held against the knife when each cut is made. In fact, the feeler wheel is *over-driven*, that is its peripheral speed is always slightly greater than the ground speed of the machine.

The weight of the feeler wheel is counter-balanced, and can be regulated to suit the coarseness of the top foliage, and the firmness of the beet in the ground. Further detail is given on this and other settings under *Operational adjustments*.

On many machines, the unit is fitted to the side of the harvester as shown in Fig 248, so that it tops one row in advance of the lifting operation.

TOP SPINNER AND TOP-SAVING DEVICES

When the tops have been severed, they must be removed from the row of beet out of the path of the lifting shares or wheels which are extracting from the soil on the subsequent bout. Fig 250 illustrates the spinner commonly used for this. It is a rotor with rubber or rubber-faced flails, which rotates at right angles to the

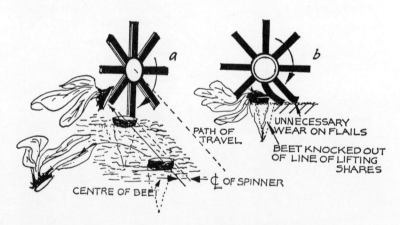

250 Setting the top spinner: *a* Alignment in relation to row— note offset. *b* Effects of spinner set too low

row of beet, sweeping the loose tops sideways on to the cleared land.

On some machines, the spinner is mounted behind the topper unit as shown in Fig 248; on others, it is fitted in front of the lifting shares so that an elevator may be fitted immediately behind the knife when the tops are to be collected directly. If the tops are required for feeding to stock, one objection to the spinner method of clearing them is the contamination with soil which inevitably results from their scattering on the ground, particularly in a wet harvest season. This renders them less digestible and palatable to stock whether used for direct feeding or for ensiling. It is for this reason that top-saving attachments are available for most complete harvesters (Fig 260). These collect the tops from immediately behind the knife and convey them to a hopper for periodic dumping, or deposit them neatly on the ground by means of a windrowing attachment. Adjustment of the discharge chute enables several rows of tops to be placed into one windrow.

BEET-LIFTING DEVICES

Various forms of beet lifters have been devised, and Fig 251 shows two types in common use at the present time. (a) shows

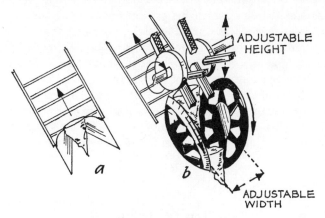

251 Beet-lifting devices: *a* Shares, *b* Ground-driven wheels and flails

a pair of triangular shares having an inclined upper surface and splayed to exert pressure on the soil at each side of the beet and so lift it, (b) shows an alternative arrangement which has become popular on its second appearance. Two slightly concave, splayed wheels are rotated by ground contact to give a more positive lifting action. When the beet is clear of the ground, rotating flails transfer it from the wheels to the elevator immediately behind. The rotation of the wheels reduces the draught of the machine, and the higher beet discharge point and corresponding elevator position results in less soil and trash being taken up the elevator. Although steering accuracy is important with either type of lifting device, it is more critical with wheels than with shares.

BEET ELEVATING AND CLEANING MECHANISM

The elevator situated behind the lifting shares or wheels is usually of the open-web or rod-link type (Fig 242). It discharges the beet either into a hopper or into an accompanying trailer (Fig 248). On the elevator, soil is knocked off the beet by the rubbing and jostling action of the rod links. In sticky soil conditions, however, this agitation is seldom sufficient. Various methods of increasing the rubbing or agitating effect are used on different machines. Sometimes the elevator chains are run on eccentric or elliptical sprockets to make them whip (Fig 244), or else a separate stationary drag chain may be draped over the moving beet (Fig 248). Another method employs a restrictor gate at the top of the elevator (Fig 252). Some machines, however, have a more elaborate mechanism in the form of revolving knockers or cages. In every case, the severity of treatment is adjustable to avoid excessive abrasion of the beet. This could be detrimental, especially if the crop has to be retained in clamp on the farm for any length of time.

BEET DISCHARGE

Beet from the harvester may be handled in various ways: by discharging from the machine into an accompanying trailer; by lifting into a large hopper for periodic discharge into a trailer

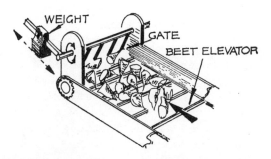

252 Beet discharge restrictor gate: adjustable resistance to regulate cleaning

by means of a second elevator; by discharging on to the ground in cross windrows; by elevating into a hopper fitted on a tractor superstructure for periodic unloading (the weight of the crop increases traction). Each method is shown in Fig 253.

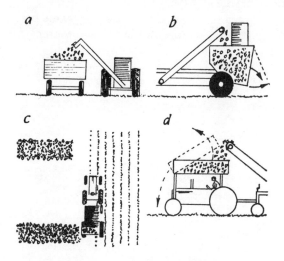

253 Alternative systems of collecting and depositing beet:
a Accompanying trailer, *b* Dumper, *c* Windrow on ground, *d* Over tractor trough

MECHANICAL PROTECTIVE DEVICES

Most harvesters are fitted with slip clutches on the main drive shaft, and on the elevator and cleaning mechanism drives. These are frequently brought into action by beet or stones blocking the mechanism, and although this may be frustrating to the operator, he should resist any temptation to over-tighten the slip clutches.

The top spinner is protected by the use of a vee-belt drive. It is important to keep this correctly adjusted.

Types of harvester

The description given above fits many current types of machine apart from variations in the disposition of elevators, types of cleaning devices, size and location of hoppers and the facilities for top saving. A simple classification of these machines is virtually impossible because many different layouts can be achieved by the interchange of attachments. However, two departures in general design and operational principle should be noted.

Fig 254 shows a machine which picks the whole beet from the ground complete with its top, and lifts it to a topping mechanism

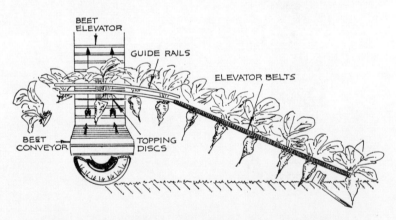

254 Side elevation of the less conventional system of lifting the beet whole and topping them afterwards

high in the machine, where it is away from the soil and stones which damage the knives of topping units mounted in the more orthodox position. The beet are loosened in the ground by a share or pair of tines, while at the same time, their tops are gripped at the base of the foliage between two rubber conveying belts which lift the entire beet to the topper unit. At the end of their travel, the beet are lined up by adjustable guides with two rotating topping discs, and as the cut is made, the roots fall into one hopper or elevator, while the tops are discharged into another. This type of machine can work at a relatively high speed, although consistency of topping depends upon reasonably strong top foliage.

Another layout is illustrated in Fig 255. In this case, the lifting

255 Principles of unorthodox beet harvester

mechanism consists of two sets of rotating lifting shares—one on either side of the row of beet. These shares knock the beet as they lift and transfer them into the discharge elevator. Good performance is claimed for this unique design.

Some makers design their harvesters as two units; the topper for attachment to the side of the tractor, and the lifting and cleaning mechanisms for attachment to the rear of the tractor. This arrangement gives the tractor driver good visibility of the topping operation, and where the lifting and cleaning assembly is mounted on the three-point linkage, its weight is transferred to the tractor and improves traction. Furthermore, the whole outfit

387

is more compact and manœuvrable than the standard trailed machine. However, attachment and detachment of the harvester is perhaps not quite so convenient where the tractor is required for transport, or for other purposes during the beet-harvesting period.

Attachment of harvester and preparation for work

The individual units of the many types of complete beet harvester have much in common, even though they may be located and disposed differently. The same basic forms of attachment exist as are common to many other types of farm machine: trailed, semi-mounted and fully mounted versions. When a harvester is attached to the tractor, the basic rules apply concerning drawbar length and power-drive shaft alignment.

Preparation tasks include lubrication as specified, checks on chain and/or belt tensions, tyre pressures, and slip clutches. The wheels on both the tractor and the harvester must be set to match the spacing of the crop rows. On machines with side-topper units, the latter must be correctly aligned (see *Operational adjustments*). The edges of the trash discs and the topper-unit knife should be sharpened, and a spare sharpened knife should be carried on the outfit. Where alternative attachments for the disposal of beet and tops are available for a particular machine, their arrangement will be determined by the system of crop collection. Where a machine is being used behind a particular tractor for the first time, a suitable ground-speed/p.t.o.-speed ratio must be attainable.

A recent development on one or two harvesters is the provision of a *live* rear axle to improve traction in difficult conditions. Some manufacturers have also developed self-propelled machines.

Operational adjustments

Operating controls on beet harvesters vary in number and type. Sometimes a second operator riding on the harvester has a steering wheel or tiller lever to keep the lifting shares or wheels astride the row of beet (Fig 256). The steering wheel or lever controls two flanged land wheels at the front of the machine, which are

388

adjustable vertically to control the depth of the lifting shares. In good working conditions, this steerage facility can be dispensed with and the swivelling wheels can usually be locked.

Other controls may include a topping unit lift-lever, and a means of raising the lifting shares at the ends of the rows. Both these controls, however, may be dispensed with on those machines which attach to the tractor three-point linkage. In this case, the

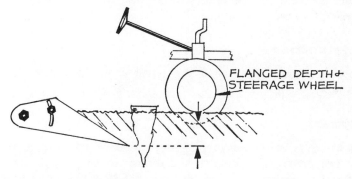

FLANGED DEPTH &
STEERAGE WHEEL

256 Setting of share depth

whole front of the machine is raised by the tractor hydraulic system. A third possibility is the provision of a remote hydraulic ram on the front of the harvester, which again is actuated by the tractor hydraulic system through an external connecting hose.

Where appropriate, control levers are provided for the discharge of beet and tops.

Independent vertical adjustment of the rear wheels of the harvester simplifies working across gradients.

There are many other operational settings beyond the scope of control levers, and some are extremely critical for satisfactory performance. A description of these, together with their possible effects on the work, follows.

TRASH DISCS

Trash discs may be located either in front of the topping unit or in front of the lifting shares to clear the machine's path of overhanging leaves (Fig 248.) The discs should be sharp and set just deep enough to cut through the leaves. On some harvesters, they are spring-loaded to follow ground contours. The angle of the discs can be varied on most machines to deflect the trash from the path of the topping knife or lifting shares. If the discs are set too high, they will be ineffective, if too low, excessive wear will result on their edges and bearings. Adjustment usually involves slackening the clamp which secures the vertical stalk of each disc, or altering the spring tension.

TOPPING UNIT

Four important adjustments are possible on the topping-unit assembly.

Alignment

On harvesters with their topping unit abreast of the lifting shares, the two processes of topping and lifting are carried out simultaneously—normally on adjacent rows of beet. The two units must therefore be spaced exactly one row width apart, and a lateral measurement should be taken at right angles to the direction of travel from a point midway between the lifting shares or wheels

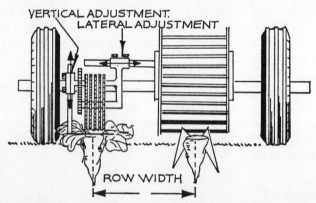

257 Alignment of topping unit to centralise on the next row to that being lifted. The vertical adjustment relates to the topping knife

to the centre of the topping knife (Fig 257). If this is different from the width between the rows of beet, adjustment can be made by slackening clamp bolts or collars and sliding the topping unit on its lateral support. Care must be taken to align the feeler-wheel drive sprockets correctly, where these are involved.

Weight relief

The weight of the topping unit is counter-balanced by a spring to relieve the beet of undue downward pressure which might push them over, or otherwise damage them. The precise amount of weight carried by the spring can be regulated by an adjustable tensioner (Fig 249). The ideal setting allows the unit to float over the beet, automatically adjusting itself to their differing heights (Fig 258). If too much weight is relieved by the balance spring, the unit will tend to bounce and give irregular topping.

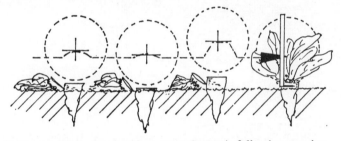

258 Rise and fall of correctly set topping unit following varying heights of beet

Generally speaking, where the tops are long and coarse, less relief is required—in other words, more weight of the topping unit should be carried by the beet to ensure penetration of the feeler wheel into the crowns. Conversely, where the foliage is delicate and the beet small or loose in the ground, more of the unit's weight should be relieved. The diameter of the feeler wheel is an important design consideration. Small-diameter wheels give more accurate topping than large ones because of their small arc of contact, but larger feeler wheels ride the beet more smoothly, and impose less forward thrust on them.

391

Vertical knife setting

The vertical clearance between the topping knife and the underside of the feeler wheel determines the amount of top removed from the beet (Fig 259). Increasing the clearance by lowering the knife will increase the amount removed; raising the knife decreases the amount removed. Removal of an unnecessary amount of top is known as *over-topping* (Fig 259a). When insufficient top is removed, leaving scars, buds and a portion of the crown at the top of the root, the machine is said to be *under-topping* (Fig 259b). A setting

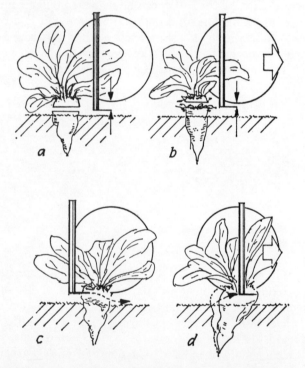

259 Topping knife adjustments: knife *a* too low (causing over-topping), *b* too high (causing under-topping), *c* too far back (causing sloping or stepped cut), *d* Too far ahead (causing beet to be pushed forward or broken off in ground)

should be obtained between these two extremes. Over-topping results in loss of sugar-producing root; under-topping results in excessive *top tare* which detracts from the farmer's financial returns. The beet have to be retopped at the sugar factory.

Longitudinal knife setting

The knife can be moved forwards or backwards relative to the centre line of the feeler wheel. This affects the timing of the cut in relation to the feeler wheel's passage over the beet. In Fig 259, c and d show the effects of the knife being *late* (too far back), and *early* (too far forward). In the first case, the feeler wheel is running off the crown of the beet as the cut is made, and in the second, the cut is started before the feeler has gained a purchase on the crown. On large diameter beet, these faults produce sloping or stepped cuts. On small diameter beet, an advanced knife pushes them forward or breaks them off in the ground. These faults are exaggerated if the topping knife is blunt.

The exact location and form of knife adjustment varies on different machines, but once the basic principles are understood, correct setting should not be difficult to achieve.

TOP SPINNER

The top spinner is usually situated immediately behind the topping unit, although two other positions are at the extreme rear of the machine in line with the topping unit, or immediately in front of the lifting mechanism, either of which leaves clearance on the side of the machine for a top-saving attachment to be fitted if required. Although the action of the spinner is extremely simple, it must be set accurately to function efficiently. If the tops are not removed cleanly, the lifting shares or wheels may become blocked. The position of the flails in relation to the beet is adjustable laterally and vertically.

Lateral setting

In order to throw the tops well clear of the roots, advantage

393

should be taken of the rising arc described by the flails as they rotate. This gives the tops a slight lifting action as the flails strike them. For this reason, the spinner is generally more effective if slightly offset in relation from the row of beet, to the side *opposite* the direction of *throw* (Fig 250a).

Vertical setting

If the spinner is set too high, the action of the flails will not be positive enough, and the top clearance will be incomplete or irregular. If the unit is set too low, on the other hand, excessive hammering of the beet and ground surface will result (Fig 250b) and if the beet are high, or loose in the soil, the impact of the flails may be sufficient to knock them out of alignment with the lifting shares which follow, or even right out of the ground. Undue pounding of the soil also accelerates wear of the flails.

TOP-SAVING DEVICE

Various patterns of top-savers have been introduced from time to time either as independent units or as an assembly incorporated in the complete harvester. The attachment may therefore be supplied either as basic equipment on the machine or as an optional extra. On most of these assemblies there is little or no adjustment, although it is sometimes possible to vary the position of the conveyor in relation to the topping knife for light or heavy top foliage, and to cope with *bolters* (beet which have run to seed and produced long, coarse stems). A choice of outlet positions from the conveyor or elevator is also provided on some machines, to facilitate delivery into a hopper, an accompanying trailer or on to a windrowing attachment (Fig 260).

BEET-LIFTING DEVICES

Shares and tines

The principle employed with twin shares is that of squeezing the beet from the ground. The inclination of the shares or tines

394

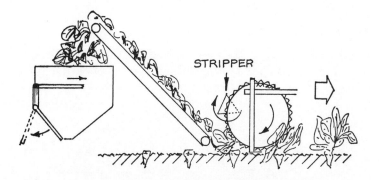

260 Top saving elevator feeding small clumping hopper

causes pressure to be exerted on the soil surrounding the beet, and this, together with the forward movement of the machine and the tapered shape of the beet, raises the latter. Three adjustments are usually possible.

Working depth

The working depth is usually controlled by depth wheels on the front of the machine (Fig 256). Excessive depth, in addition to imposing unnecessary draught on the outfit, detracts from the harvester's cleaning efficiency especially in sticky soils. On stony land, it may also increase the incidence of blockage. Insufficient depth, on the other hand, may result in beet being either broken off or left whole in the ground. Scrapers are generally fitted on the depth wheels and these should be set to keep the rims clear of soil and the depth thereby constant.

Pitch

The pitch is the angle of inclination of the shares (Fig 261). The extremes of this setting can produce the same effects as extremes of depth adjustment. Ideally, the pitch should be the minimum necessary for clean lifting.

Width

Width refers to the clearance allowed between the two shares (Figs 251 and 261). The best setting depends upon the average diameter of the beet. If there is too much clearance between the

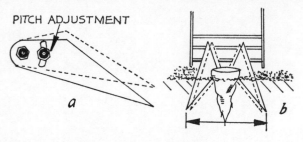

261 Setting of lifting shares: *a* Pitch, *b* Width

rear of the shares, small beet may pass through them and so miss the elevator. If the shares are too close together, excessive quantities of soil and stones may be taken up the elevator.

Squeeze wheels

The squeeze-wheel type of lift mechanism employs the same principle of side pressure on the beet as fixed shares, but there is also a lifting action by the rising rear portions of the wheels as they rotate by ground contact (Fig 251b). The beet are discharged from the wheels on to the elevator by means of rubber-faced flails. The advantages claimed for this system of root lifting are reduced draught, less likelihood of blockage and broken beet, and improved cleaning of the beet. The higher intake point of the beet elevator, which is generally possible with this lifting mechanism, reduces the intake of soil, stones and trash. However, the rotation of the lift wheels makes them more dependent on accurate steering than the static shares, and the diameter of the wheels is an important design factor. With their basic setting, the wheels are capable of handling beet of varying size, although width adjustment is possible by means of spacers on their spindles. The flails or beaters which transfer the beet from the wheels to the bottom of the elevator can also be adjusted—a high setting being necessary for large beet, and a low position for smaller beet.

CLEANING MECHANISM

On most complete harvesters, the beet are cleaned as they are being elevated. Open-web elevators rub and agitate the beet in transit, and the intensity of the agitation may be regulated in various ways. Eccentric jockey sprockets (Fig 244), weighted drag chains (Fig 248), or discharge-restriction gates (Fig 252) are some of the means provided. A few machines employ a special cleaning device in the form of a cage having internal projections to tumble and jostle the beet. Whatever form the cleaning mechanism takes, adjustments are usually simple, and the objective remains the same. This is to remove as much soil as possible without damaging the beet. Damage must be avoided, particularly if the crop is to be clamped for storing on the farm for any length of time before going to the sugar factory. This may well be the situation on heavy land farms, where it may be prudent to lift the crop early before bad weather sets in. Beet sometimes have to be held by the farmer pending delivery to the factory—a matter controlled by permits.

Special attachments

Special attachments include the following:
1. Top-slasher. This is an attachment fitted in front of the topper unit (or one row ahead) to cut down coarse bolted tops (Fig 262). On some farms, however, bolted tops are first cut down with a flail harvester.
2. Top-saver. This may take one of several forms, but the most common is simply an elevator situated immediately behind the topper unit. As described previously, this conveys the tops to a windrower or dumping hopper.
3. Alternative types of lifting shares. Interchange of shares or squeeze wheels is sometimes desirable to suit different soils and working conditions.

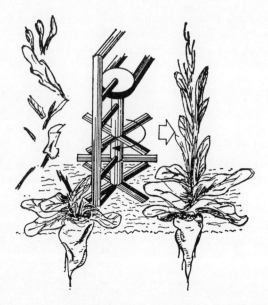

262 Top-slashing attachments for destruction of bolted tops

4. Beet discharge elevator and/or dumper. Most of the arrangements previously described can be achieved by the fitting of
these optional assemblies.
5. Carrot attachment. The harvesting of carrots requires a more
delicate topping action, and a different type of lifting share, and
the necessary components for this adaptation are available for
several sugar-beet harvesters.

Field procedure

The efficiency of a beet harvester is influenced by many factors.
Apart from the more obvious ones of soil condition, weather
and bulk of crop, certain factors are determined long before
harvest time. Indeed, they should receive careful consideration when first planning the crop. Brief reference to these
follows:

Levelness of seedbed

The importance of a level field surface has been mentioned in connection with row-crop cultivation (Chapter 2). A level surface is equally important for accurate topping and lifting of beet at harvest time. One-way or square ploughing help to achieve levelness, but even so all cultivations should be performed with this important objective in mind.

Choice of seed and time of drilling

The seed selected and the time chosen for drilling can have an important bearing on the number of bolters present at harvest time. In addition to the problems these can cause in the harvesting operation, there is also a considerable loss of sugar content in the affected beet.

Row width

A row spacing of 20 in is now common practice for the sugar-beet crop. This gives reasonable space for the tyres of standard general-purpose tractors when used for inter-row cultivations and for mechanical harvesting. Some growers, however, prefer other row spacings for their required plant population, and special row-crop and other narrow-wheeled tractors are available to suit the narrower rows. The draught of a complete harvester, on heavy land and in wet conditions, can be very high and may well be beyond the scope of lighter tractors.

System and accuracy of drilling

An adequate headland should be provided to facilitate entry and exit of the tractor and harvester at bout ends. On undulating land, the drilling should be carried out in the direction likely to be most favourable for the harvesting operation—both for the lifting and the collection of the crop. The drilling bouts must be carefully matched to one another, especially where the harvester to be used has the more usual side-mounted topping unit, as occassionally the topping unit and the lifting mechanism may have to straddle a joint in the drilling bouts. This will inevitably happen whenever an odd number of seeder units is used on a tool bar for the drilling operation.

Hoeing and thinning

A lot of weeds and an irregular crop can affect the performance of the beet harvester to some degree.

It should be clear that the efficiency of the harvester depends upon good management and operational skill at every stage of the crop's production.

OPENING OUT THE FIELD

Where the headland has been fully planted, it is usual to lift the beet from the corners by hand. With machines having an abreast topping unit, the initial circuit must be made around the field with the lifting mechanism raised out of operation. Assuming a machine to have a right-hand topping unit, the first run may be made in an anti-clockwise direction—topping the outer row. At the end of this circuit the lifting shares can be lowered and the outfit may proceed clockwise until the headland has been completed. Alternatively, the first circuit may be made in a clockwise direction, topping the second row in from the boundary, and the reverse circuit later when more room is available to turn the outfit.

When the headland is clear, the straight rows are worked out in lands.

OPERATING TECHNIQUE

The manufacturer's recommendations, concerning the running speed of the machine, should be followed. Too high a speed can damage the mechanism, while inadequate speed can seriously affect the output. As the elevators and conveyors maintain a constant speed, the ground speed must be carefully regulated to suit the density of the crop. In good working conditions, the topping unit may well have a capacity in excess of the elevating mechanism, and in such circumstances, the tractor driver is often tempted to accelerate ground speed without realising that he is overloading the back half of his machine—that is until a slip clutch protests. Ground speed may be further restricted when the beet are loose in the soil, when they are very irregular in size and spacing, or when there is an excessive amount of trash or high proportion of bolters. The use of a top-slashing attachment already described is a great help when faced with bolted beet.

If the topping unit or lifting shares are blocked, clearing is simplified by lifting the appropriate assembly out of work. Those machines which attach to the three-point linkage of the tractor have an advantage in this respect.

It is usual, with most side-topping harvesters, to top the beet immediately adjacent to the row being lifted as shown in Fig 263.

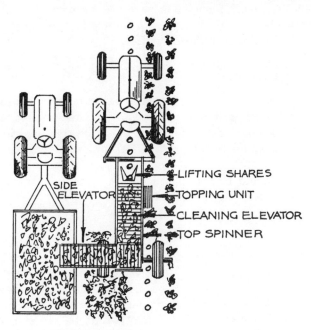

263 Plan view of harvester at work

In some cases, however, the topping unit has sufficient lateral adjustment to position it *two* rows ahead of the lifting shares instead of one, and this is sometimes done when tops are required in a clean condition for feeding, as by this method the tops are discharged in windrows clear of the path of accompanying tractors and beet trailers, thus avoiding fouling of the tops with soil.

OPERATIONAL CHECKS

The following sequence may be followed for checking the performance of a complete sugar-beet harvester.

Accuracy of topping

Reference has been made to over-topping and under-topping. Over-topping indicates too low a knife setting, and under-topping too high a setting. Although it is inevitable that a small proportion of beet will be imperfectly topped, it is important to aim at a high standard of accuracy in this operation.

Squareness of topping

The squareness of the topping may be upset by a number of factors: (a) incorrect timing of the knife, i.e. the knife is too far forward or too far back, (b) blunt knife—in this case the impact of the knife forces the root forward as the cut is made, (c) excessive ground speed, (d) wet soil or loose beet—in these circumstances a keen edge should be maintained on the topping knife, and the topper balance spring carefully adjusted.

Consistency of topping

Even topping may be affected by irregular beet spacing, the height and the development of the crop. Such conditions are probably the most difficult to cope with, and the best action to take is to reduce ground speed and set all the adjustments on the topper unit carefully, giving special attention to the balance-spring adjustment.

402

Top clearance

The setting of the top spinner can be a problem where heavy, coarse foliage has to be handled. Incomplete clearance of the severed tops from the beet may be due to too high a setting or incorrect lateral position of the spinner in relation to the row of beet. Some experimenting is usually necessary for the best effect, and it may help to offset the top spinner as described under *Operational adjustments*, although the desirable amount of offset varies from one crop to another.

If the beet are displaced from their row or lifted from the ground completely by the spinner, the setting is too low. This can damage the flails as well as displace the beet. On some machines, the tops are collected from behind the knife on the elevator of a top-saving attachment, and adjustments are sometimes possible to regulate the clearance between the knife and elevator.

Lifting of beet

If problems are encountered in the lifting of the beet, the following possible causes should be looked for: (a) lifting shares or wheels not deep enough, (b) shares too wide, (c) pitch of shares too shallow, (d) shares misaligned or driving inaccurate, (e) shares obstructed, preventing full penetration.

If the beet break off in the ground, this may be due to excessive pitch or too shallow a setting of the lifting shares. On occasions, the lifting of beet can be hampered by the hardness of the soil.

Cleaning of beet

The beet is cleaned by rubbing or impact with some form of knocker, shaker, elliptical chain sprockets, or restricted elevator outlet as previously described. As the intensity of cleaning required is always dependent upon the prevailing soil conditions, the cleaning mechanism should be regulated to achieve the highest degree of cleaning possible without abrasing or bruising the beet.

Where the cleaning intensity is regulated by restricting the outlet of the crop from conveyors or elevators, the adjustment inevitably affects the throughput of a machine, and better cleaning may impose a lower ground speed.

Most of the harvester's components are subjected to considerable strain and vibration, and so it is important to ensure that all adjustments are firmly secured once the correct settings have been achieved.

CROP-HANDLING SYSTEMS

The various systems of beet handling during the harvesting operation are: (a) discharge of beet direct from lifting and cleaning mechanism to an accompanying trailer by means of a side-delivery or rear-delivery elevator, (b) retention of lifted beet on the machine in a large-capacity hopper for discharge at intervals either into a transport vehicle or on to the ground at the headlands or clamping site, (c) discharge of beet direct from lifting and cleaning mechanism on to the ground in heaps by means of a dumping attachment in such a way as to form cross windrows for later collection, (d) employment of a front-delivery elevator on the harvester which conveys the beet from the cleaning mechanism into a container carried on a superstructure above the tractor. At intervals the beet is discharged from the container by means of a tipping mechanism. The last method is not so widely used.

The tops may be discharged by scattering over the ground surface by the spinner, or by discharge on to the ground either with 'in line' windrows—two or three rows in one—by means of a continuous windrowing attachment, or in cross windrows by means of a dumping attachment.

Maintenance

DAILY ATTENTION

The widely varying types of bearings used in the construction of modern beet harvesters makes servicing recommendations rather specific to each make of machine. Increasing use is being made of sealed, ball and roller bearings which give long periods of service between greasings. However, there are usually some simple bearings which require fairly frequent attention, but as with all machines, the maker's recommendations should be followed. Roller chains which are protected from grit and dust should be lubricated, but those in exposed positions are usually best left dry. Powdered graphite might be used, but this is rather expensive.

Although not usually specified as a daily routine job, the topping knife may require daily sharpening in some field conditions, particularly where large stones are present on the surface. Some topping knives are reversible, and provide two periods of service between sharpenings. When sharpening knives on a power grinder, avoid overheating the edge, or the steel will be softened.

PERIODIC ATTENTION

Other maintenance checks and adjustments necessary at intervals include: tyre pressures, chain tensions and/or vee-belt tensions. The condition of the top-spinner flails should also be checked; if one is lost, the rotor will be thrown out of balance, and vibration may damage its bearing. The trash discs should be sharpened when they become ineffective, especially when working in soft, wet, soil conditions.

END-OF-SEASON ATTENTION

In addition to the many strains and stresses imposed upon the beet harvester when at work, considerable deterioration can follow during its period of idleness, unless it is properly cleaned, checked over and housed until next required. The following points are some of the more important ones:

1. The machine should be thoroughly cleaned down and protected with rust preventive. Where a pressure hose or steam cleaner is used, care should be taken to avoid forcing water or steam into sealed bearings, especially when chemical detergents are added.

2. The whole machine should be greased, and while this is being done checks may be made for excessive wear on the various bearings.

3. The chain links, sprockets and idlers should be closely inspected for wear and, if necessary, arrangements put in hand for their reconditioning or replacement.

4. Slip clutches which have been subjected to repeated operation should be dismantled and inspected for wear on their elements.

5. Any vee belts should be removed, examined and stored away in a dark place. The pulley faces should be protected against rust.

6. As with other seasonal machines, it is good practice to relieve the tyres of the machine's weight.

7. The power-drive shaft universals and the bearings of the primary shaft on the machine should be examined.

REPLACEMENT PARTS

In so far as they are applicable, it is advisable to have the following spares on hand at the commencement of a season's work: two topping knives, a set of top-spinner flails and their necessary retaining pins or bolts, lifting shares, slip-clutch elements and springs, chain links and vee belts.

Safety precautions

The sugar-beet harvester is a machine which has to perform relatively delicate operations in extremely arduous conditions. Blockages can be frequent, and it is not always easy to clear them. Furthermore, even the tidiest design of harvester is still a relatively ungainly piece of machinery. These factors have contributed to many minor injuries (and some serious ones) being suffered by operators. The warnings against interfering with the machine when its mechanism is in motion, and failure to replace guards, cannot be overstressed. As the machine is often operated by two men, care is also needed to ensure that the mechanism is not put in motion by one operator, whilst the other is making adjustments.